Contract Pro
for Surveyors

Contract Practice
for Surveyors

THIRD EDITION

Jack Ramus, ARICS

AND

Simon Birchall

BSc(Hons) ARICS MACostE PGCE

Laxton's
An imprint of Butterworth-Heinemann
Linacre House, Jordan Hill, Oxford OX2 8DP
225 Wildwood Avenue, Woburn, MA 01801-2041
A division of Reed Educational and Professional Publishing Ltd

R A member of the Reed Elsevier plc group

OXFORD AUCKLAND BOSTON
JOHANNESBURG MELBOURNE NEW DELHI

First published as *Contract Practice for Quantity Surveyors* 1981
Reprinted 1982
Second edition 1989
Reprinted 1990, 1991, 1993, 1994
Third edition 1996
Reprinted 1998, 1999, 2001 (twice)

British Library Cataloguing in Publication Data
Ramus, J. W.
    Contract practice for surveyors - 3rd ed.
    1. Construction contracts - Great Britain
    I. Title   II. Birchall, Simon   III. Contract practice for quantity
    surveyors
    692.8'0941

ISBN 0 7506 2661 5

For more information on all Butterworth-Heinemann
publications please visit our website at www.bh.com

Composition by Genesis Typesetting, Rochester, Kent
Printed and bound in Great Britain by Biddles Ltd
www.biddles.co.uk

Contents

Figures and tables

Tables

Preface

Further changes in contract practice, legislation and the ways in which construction is procured and financed have occurred since the second edition of this book was published in 1989. The seventh edition of the *Standard Method of Measurement of Building Work*, published in 1988, is now well established. It was not practicable to apply its provisions in the examples for the reason stated in the preface to the second edition, but this has been done in this new edition. Further amendments to the *JCT Standard Form of Building Contract 1980* have also been taken into account.

The scope of the book has been widened by the addition of chapters on CDM Regulations, insurance, taxation, collateral warranties, guarantees and bonds, insolvency, disputes, and contract selection. Because of this, and because we feel that the book will be useful to surveyors across a wider spectrum of the construction field, particularly building surveyors and general practice surveyors, the title has been changed to reflect this.

The book provides a useful reference for degree students, graduates, practitioners engaged in all aspects of construction contracts, and candidates sitting membership examinations of the various construction industry professional associations. We hope that the detailed examples will continue to meet the need for clear guidance in the practical procedures relating to contract practice.

We emphasize again that much of the text is applicable to the duties of surveyors employed in building contracting companies, although it has been written principally for those employed in private practice or in central or local government departments, and attention has been given to those duties where they differ from private practice. Throughout the book, where specific reference is made to surveyors in construction companies, the term 'contractor's surveyor' is used; where reference is to surveyors in general, the term 'the surveyor' is used.

Attention is drawn to the fact that the text relates only to English law and no reference is made to differences in law and practice applicable in Scotland. It is hoped nonetheless that Scottish surveyors will find the book of use and value in their studies and/or employment.

We wish to acknowledge the help and encouragement we have received in the preparation of this third edition. We thank RIBA Publications Ltd for permission to reproduce extracts from the JCT Standard Form of Building Contract and to quote from others of their publications; to the Royal Institution of Chartered Surveyors for permission to reproduce standard forms of valuation; to the Department of Health for allowing us to use their expenditure forecasting formula; and to the Working Group on Indices at the Department of the Environment for permitting the use of material from their publications relating to Price Adjustment Formulae.

Jack Ramus
Simon Birchall
March 1996

1

Contracts and the surveyor

The nature and form of contracts

All aspects of the surveyor's professional work relate directly or indirectly to the costs of construction work of all kinds. This includes new building and civil engineering projects, and the alteration, extension or refurbishment of existing buildings. In addition, his expertise extends to consequential costs incurred by owners and occupiers after the completion of construction projects. These include repairs and redecoration, replacement of components and the operating costs of installations such as lifts, air-conditioning, heating, etc. All these costs are known collectively as 'costs-in-use' or 'life cycle costs'.

Such costs are incurred when building and engineering contractors of all kinds carry out work for clients of the construction industry. This necessarily requires that the parties concerned in a particular project enter into a contract, i.e. a legal agreement to discharge certain obligations. So, as a simple example, a builder enters into a contract with the owner of a plot of land in which he undertakes to build a house complying with the owner's requirements. On his part, the owner undertakes to pay the builder a stated sum of money.

In its simplest form a contract may be an oral agreement, the subsequent actions of the parties providing evidence of the existence of a contract. Such an agreement is legally binding, provided it does not entail an illegal act.

Alternatively, a contract may be set out in a letter or other brief document stating the essential matters agreed upon. However, such forms of contract are only suitable for the simplest projects, such as building a garage onto an existing house or erecting fencing around a plot of land. Even in those cases disputes may arise, leading to arbitration or litigation, both costly processes.

In the case of most building and engineering projects therefore it is essential that the conditions under which a contract is to be carried out and the obligations, rights and liabilities of the parties should be fully and clearly stated in writing. Even then, disputes can still arise because of ambiguity or omission in the contract documents or because something unforeseen arises during the period of the contract.

For most projects, printed standard forms are used as the basis of the contract. These set out the conditions applicable to the contract and cover all the matters which experience has shown need to be specified in order that both parties may have a clear picture of their rights and obligations.

Those matters which are peculiar to a particular project, such as names of the parties, the contract sum, commencing and completion dates, etc. are incorporated into the contract by insertion into blank spaces in the Articles of Agreement or the Appendix, both of which are part of the printed document. It is sometimes necessary for the printed text of a clause of a standard form to be amended, added to or deleted in order to accommodate particular requirements of a building owner or because the standard wording does not suit the particular circumstances of a project. Such amendments must be made with care, to avoid introducing ambiguities or inconsistencies with other clauses or other contract documents.

The National Joint Consultative Committee for Building (NJCC) has issued a Procedure Note[1] advising against extensive amendments to the JCT Form, on the grounds that such amendments may introduce inequity and that they tend away from greater standardization of building procedures. The Note also advises that when amendments are considered necessary, only a competent person should amend the Form so that ambiguities or inconsistencies which might lead to difficulties later are not introduced. This advice applies equally to all standard forms.

The advantages of using standard forms of contract instead of composing a contract document afresh for each project are:

(a) because they are written by legal experts, ambiguities and inconsistencies are reduced to a minimum;

(*b*) the rights and obligations of both parties are set out clearly
and to the required degree of detail;

(*c*) because of standardization, frequent users are familiar with
the provisions in a particular Form, resulting in a greater
degree of consistency in their application;

(*d*) the time and expense which would otherwise be incurred in
preparing a fresh document for each occasion is avoided.

The standard forms most widely used for building projects are
those in the range produced by the Joint Contracts Tribunal
(JCT), a body which is composed of representatives of the
following organizations:

Royal Institute of British Architects
Building Employers Confederation
Royal Institution of Chartered Surveyors
Association of County Councils
Association of Metropolitan Authorities
Association of District Councils
Confederation of Associations of Specialist Engineering
    Contractors
Federation of Associations of Specialists and Sub-
    Contractors
Association of Consulting Engineers
British Property Federation
Scottish Building Contract Committee

A list of the JCT Standard Forms of Contract and Sub-Contract is
given below:

1   Standard Form of Building Contract 1980 Edition (JCT 80)
    (*a*) Private With Quantities.
    (*b*) Private With Approximate Quantities.
    (*c*) Private Without Quantities.
    (*d*) Local Authorities With Quantities.
    (*e*) Local Authorities With Approximate Quantities.
    (*f*) Local Authorities Without Quantities.
2   Standard Form of Building Contract With Contractor Design
    1981 Edition (CD 81).
3   Intermediate Form of Building Contract 1984 Edition (IFC
    84).

4   Standard Form of Management Contract 1987 Edition (MC 87).
5   Standard Form of Prime Cost Contract (PCC 92)
6   Agreement for Minor Building Works (MW 80) 1980.
7   Measured Term Contract (JCT 89).
8   Standard Form of Nominated Sub-Contract (NSC/A).
9   Sub-Contract Conditions to IFC 84 (NAM/SC).

The above Standard Forms of Contract are dealt with in detail in Chapter 22, *Contract selection*.

In addition, there are various amendments, supplements and Practice Notes relating to JCT Forms. A complete catalogue of JCT publications is available from RIBA Publications Ltd.

Other standard forms of contract are:

(a) ICE Conditions of Contract (for civil engineering works, published by the Institution of Civil Engineers and the Federation of Civil Engineering Contractors).
(b) General Conditions of Government Contracts for Building and Civil Engineering Works (known as GC/Works/1 and published by Her Majesty's Stationery Office).
(c) ACA Form of Building Agreement (published by the Architectural Press Ltd for the Association of Consultant Architects).
(d) ACA Form of Building Agreement 1984, BPF Edition (published by the British Property Federation Ltd).

The above list is not exhaustive. See Chapter 22, *Contract selection*, for further information about non-JCT Forms of Contract.

The surveyor's role

Many of the provisions in standard forms of contract relate to financial matters, and surveyors need to have a detailed knowledge and understanding of them and be able to apply those provisions properly and in a professional manner. Consequently, most surveyors have acquired considerable expertise in the use of standard forms of contract and many architects look to the surveyors with whom they regularly work for advice and guidance on contractual matters. Frequently it falls to the surveyor to prepare the contract documents ready for the execution of the contract by the parties, a duty which requires particular care and attention to detail.

In the performance of his duties during and after the construction period of the project, the surveyor has the duty to ensure that all actions taken in relation to the financial administration of the contract will be fair to both parties.

Whilst there often is, and should be, such an attitude underlying the surveyor's approach to his duties, he must not forget two important facts. First, 'fairness' is often a subjective matter so that an action or decision may be seen as fair by one person and as very unfair by another. Secondly, the surveyor's authority to act rests ultimately upon the terms of the contract and, consequently, his actions must always be within the scope of those terms. Both parties to a contract, Employer and contractor, are bound by the terms of that contract, having been freely agreed to by them when they signed the contract. Should any one of those terms in its practical application become disadvantageous, or even financially disastrous, to either party, the surveyor has no authority to vary its provisions in order to alleviate those consequences.

The young surveyor may be tempted, by a sense of fair play which in other circumstances would be very commendable, to act in a particular situation in a way which seems to him to be fairer to one or other of the parties rather than in accordance with the strict application of the contract terms. For example, it might seem very unfair that a contractor, having complied with a verbal instruction by the architect to carry out a variation to the design of a building, should be refused payment for the extra cost just because the variation had not been confirmed in writing. The surveyor cannot do otherwise, however unfair it may seem, if the contract requires that all variations, to be valid, must be in writing.[2] Similarly, if the surveyor happens to know during the construction period that either party is having cash flow problems, that knowledge must not be allowed to influence the value he puts upon work completed or unfixed materials on site when preparing a valuation for an interim certificate. However sympathetic he may feel towards either party, the surveyor must not allow his personal feelings to prevail over his judgement as to the proper application of any term of a contract. He must remember that any unwarranted advantage given to one party is likely to be at the same time an equal disadvantage to the other party. The foregoing remarks apply particularly to surveyors employed in private professional firms and in the offices of local authorities, government departments, and other public bodies.

The surveyor employed by a building contractor is in a somewhat different situation in that his responsibility and loyalty is primarily to his employer. Nevertheless, his actions on behalf of the contractor must equally be governed by the terms of each contract with which he deals. There is no reason, however, why the contractor's surveyor should not invoke a term of a contract to the advantage of his employer and there can be no objection to his so doing. For example, by a careful comparison of the bills of quantities and the drawings, he may be able to detect errors of omission in the quantities, the correction of which may be provided for in the contract.[3] On the other hand, he must recognize that a claim, of whatever kind, can only be justified if it is based upon a specific term of the contract and, even if shown to be valid, the valuation of it may only be reached by prescribed means.[4]

But in whatever kind of organization a surveyor works, his attitude to the carrying out of his duties should be a completely professional one. That is to say, the quality of the work he does should be as high as he is able to achieve, without becoming pedantic. Despite the pressures of a highly competitive business world, as a member of a respected and worthy profession, the surveyor should aim to be entirely honest and impartial in his interpretation and application of the terms of a building contract, on whichever side of the contractual 'fence' he may be.

The surveyor's post-contract duties

The following is a summary of the duties normally carried out by the surveyor following acceptance of a tender for a project where an architect is responsible for the supervision and management of the contract. Some of course may not be applicable to particular projects. The respective chapters in which those duties are described in detail are given in brackets.

Before work starts on site:

1 Arrange for contract documents to be prepared ready for signatures of the parties (Chapter 6).

2 Prepare forecast of 'rate of spend' during the construction period and advise client on anticipated liability for payments on account to the contractor, giving dates and amounts (Chapter 14).

3 Make preliminary arrangements for preparing valuations for payments on account in consultation with the contractor's surveyor; analyse Preliminaries and calculate amounts of time-related payments and percentage rate of cost-related payments; prepare schedule for stage payments, if applicable (Chapter 12).

During construction period:

4 Prepare valuations for payments on account at the intervals stated in the contract and agree with contractor's surveyor (Chapter 12).

5 Plot payments on account on 'rate of spend' graph and report to architect on any significant divergences (Chapter 14).

6 Prepare estimates of likely cost of variations on receipt of copies of architect's instructions; later measure and value; check and price daywork vouchers (Chapter 8).

7 Advise architect, if requested, on expenditure of prime cost sums; check nominated sub-contractors' and nominated suppliers' final accounts and adjust contract sum accordingly (Chapter 7).

8 Advise architect, if requested, on expenditure of provisional sums; measure and value work carried out by the main contractor against provisional sums (except where lump sum quotations have been accepted) and adjust contract sum accordingly (Chapter 7).

9 Prepare financial reports for architect and client at the time of interim payments (Chapter 14).

10 Check main contractor's claims for increases in costs of labour, materials, levies, contributions and taxes, etc., if applicable; alternatively, apply price adjustment indices to amounts included in interim valuations (Chapters 9 and 10).

11   Measure projects based on schedules of rates or on bills of approximate quantities as the work proceeds, either on site or from architect's drawings, and value at contract rates.

12   Advise architect, if requested, on contractor's claims (if any) for loss and expense payments; if accepted, negotiate claims with contractor (Chapter 11).

After construction period:

13   Prepare final account and agree details and total with contractor's surveyor (Chapter 13).

The surveyor's legal obligations

legal status in terms of JCT 98

The surveyor is normally appointed direct by the client and a contract of employment is entered into by the two of them. This may be in the form of a letter setting out the essential terms of the contract, such as the services to be provided and the basis of payment. Often the Form of Agreement for the Appointment of a Quantity Surveyor, published by the Royal Institution of Chartered Surveyors, is used.

The obligations of the surveyor are governed therefore by the terms of the contract of employment. However, a person who offers professional services has a duty in law to exercise proper skill and care in the performance of those services, otherwise he may be liable for damages under the law relating to negligence. His actions are expected to be of an equal standard to those of other qualified persons practising the same profession.

Only a few cases regarding surveyors have come before the courts, and the following are of particular interest.

London School Board v Northcroft Son & Neighbour (1889)

The clients brought an action against Northcrofts, the quantity surveyors, for negligence because of clerical errors in calculations which resulted in overpayments to the contractor. It was held that as the quantity surveyor had employed a skilled clerk who had carried out a large number of calculations correctly, the quantity surveyor was not liable.

Dudley Corporation v Parsons & Morris (1959)

This case centred around the interpretation of an item in a bill of quantities which stated 'extra over for excavating in rock'. It was unclear whether the item referred to the immediately preceding item of basement excavation or to all the excavation items. The seriousness of the matter from the contractor's point of view was that the item had been grossly underpriced and on remeasurement was some three times the provisional quantity in the bill. It was held by the Court of Appeal that the contractor was only entitled to be paid at the erroneous rate for the total quantity of the item as remeasured.

Tyrer v District Auditor of Monmouthshire (1974) 230 EG 973

The local authority overpaid a contractor because Tyrer, a quantity surveyor who was an employee of the authority, had accepted rates for work which he must have known were ridiculously high and had also made an arithmetical error when issuing an interim certificate. Tyrer appealed against being surcharged by the District Auditor for the loss sustained by the authority, on the grounds that he was acting in a quasi-judicial position. The appeal was rejected. The quantity surveyor owed a duty to carry out his professional work with a reasonable degree of care and skill.

Sutcliffe v Thackrah (1974) HL 1 Ll Rep 318

Although in this case the defendant was an architect, the quantity surveyor had included in two valuations the value of work which was defective, not having been advised of the defects by the architect. The client sued the architect for the cost of rectifying the defective work. The House of Lords held that the normal rules of professional negligence applied to all aspects of an architect's duties and, in exercising his functions, he must act impartially. The same obligations also apply to the actions of a quantity surveyor, although that was not a matter under consideration in this case.

John Laing Construction Ltd v County and District Properties Ltd (1982) QB 23 BLR 1

In this case (concerning a contract based on the Standard Form of Building Contract, 1963 edition, July 1977 revision) the quantity surveyor had included in interim certificates fluctuations amounts payable to a sub-contractor which ought not to have been included because the written notice required in clause 31D(1) had not been given.

The judge concluded that clause 31D(3) of the contract conferred upon the quantity surveyor 'no authority to agree, or to do anything which could have the effect of agreeing, liability as distinct from quantum'. However, he said that 'a quantity surveyor appointed under this or any other contract can as a matter of fact and law be given such authority, or any other authority, by the Employer in any one or more of a number of ways'. The Employer was entitled to have the sums in question excluded from the final account.

References

1. NJCC, *Procedure Note No. 2, Alterations to Standard Form of Building Contract* (London: NJCC, 1981).
2. JCT, *Standard Form of Building Contract, 1980 Edition* (London: RIBA Publications Ltd, 1988), clause 4.3.1.
3. ibid., clause 2.2.2.2.
4. ibid., clause 13.5.

Bibliography

1. CECIL, Ray, *Professional Liability* (London: Legal Studies & Services 1991).

2

Building procurement – traditional methods

From early in the nineteenth century until about the 1950s, the ways by which building projects were promoted and carried out in the UK conformed to straightforward and well-tried procedures. If the project was small, the building owner (or 'Employer', as he is often called) employed a building contractor to design and construct the building for him. Because buildings generally conformed to a well-defined pattern, contractors had within their organizations the full range of expertise and skills normally required.

In the case of larger projects the Employer appointed an architect to design the building, and he then produced drawings and a specification. If the architect considered it necessary (and the Employer approved), he then appointed a quantity surveyor to prepare a bill of quantities. Then, on the basis of either the specification and drawings or the bill, contractors were invited to tender in competition to carry out the work. Usually the lowest tenderer was awarded the contract.

Since the mid-1940s the architect's nomination and/or appointment of the quantity surveyor has been gradually superseded by direct appointment by the Employer, sometimes before the selection of the architect and, in some cases, the latter's selection is made on the recommendation of the quantity surveyor.

The traditional methods of building procurement are still widely used and their respective distinguishing characteristics are as described below. They have gradually evolved during the

twentieth century to meet changing circumstances and techno-
logical developments, and variants of the main procedures have
been introduced from time to time, but the essential principles
still apply.

A. Based on bills of firm quantities

The building owner commissions an architect to prepare a design
and, upon virtual completion of the design, the surveyor prepares
a bill of quantities based upon the architect's drawings and
specification information. Contractors are invited to price the bill
and submit tenders in competition for carrying out the work. The
contractor submitting the lowest tender is usually awarded the
contract.*

The essential characteristics of this method are (i) that both the
quantities and the unit rates in the bill form part of the contract
and (ii) that virtual completion of the design precedes the signing
of the contract. Such a contract is a *lump sum contract*
(sometimes called a *fixed price contract*) because a price is stated
in the contract as payment for the work described in the bill.

Advantages

1   Both parties have a clear picture of the extent of their
    respective commitments.
2   The unit rates in the bills provide a sound basis for the
    valuation of any variations to the design.
3   A detailed breakdown of the tender sum is readily available.

Disadvantages

1   The length of time taken in the design of the project and in the
    preparation of the bills of quantities.
2   The problem of dealing with those variations which are so
    fundamental or extensive as to change the character of the
    remainder of the work or the conditions under which it has to
    be carried out.

---

* The JCT Standard Form of Building Contract With Quantities, 1980 Edition
  (Private or Local Authorities version as appropriate) is commonly used.

B. Based on bills of approximate quantities

This method is largely similar to the preceding one, except that the quantities given in the bill are approximate only and are subject to later adjustment. The essential characteristics are (i) that only the unit rates form part of the contract, and (ii) the signing of the contract and the beginning of work on site may proceed before the design is complete.

The bill of quantities is normally specially prepared for the particular project and descriptions of work are as detailed as in a bill of firm quantities, but the time otherwise required for detailed measurement of the quantities is saved, the quantities given being estimates of likely requirements. Sometimes the bill re-uses the quantities which were prepared for an earlier project of a sufficiently similar kind and size.

This method results in a contract* which is sometimes regarded as a lump sum contract although it is not strictly so, there being no total price stated in the contract. In effect, it is very similar to a schedule of rates contract (see D below).

Advantages

1 Construction on site may begin earlier.
2 The extra expense of preparing firm quantities is avoided (although this is offset by the cost of fully measuring the work as actually carried out).

Disadvantages

1 The bills of quantities cannot be relied upon as giving a realistic total cost at tender stage and in consequence, the parties to the contract are less certain of the extent of their commitment.
2 The construction works have to be measured completely as actually carried out, which may prove more costly than to have prepared bills of firm quantities initially.
3 The architect may feel less pressure to make design decisions which ought to be taken at an early stage.

---

* The JCT Standard Form of Building Contract With Approximate Quantities 1980 Edition (Private or Local Authorities version as appropriate) is commonly used.

C. Based on drawings and specification

This method closely resembles that described in A above, the difference being that no bills of quantities are supplied to tenderers, who have to prepare their own quantities from the drawings provided. This procedure is intended to be used for relatively small works (say, not exceeding £50,000) and for sub-contract works, although it is not unknown for quite large contracts to be tendered for on this basis. A survey carried out by the Junior Organization of the RICS showed that the average value of contracts let on this basis during 1987 was just under £200,000.[1]

The essential characteristics are (i) that tenderers are supplied only with complete working drawings and a full specification, and (ii) that virtual completion of the design must precede the signing of the contract.*

Advantages

1   The time required for the preparation of tender documents is reduced, as the time-consuming process of preparing bills of quantities is eliminated.
2   Both parties can have a clear picture of their respective commitments at the time of signing the contract.

Disadvantages

1   No breakdown of the tender sum is immediately available (although the tenderers may be asked to provide a Contract Sum Analysis, either as a part of their tender submission or subsequently).
2   The valuation of variations presents problems, as indicated above.
3   There is little, if any, control over the percentage rates for additions for overheads and profit to the prime cost of labour, materials and plant elements in dayworks (defined on p. 114). Tenderers are normally asked to state percentage rates to be used in the event of dayworks arising. Where such rates have no effect on the tender sum, there is little incentive to the tenderer to moderate them.

---

* The JCT Standard Form of Building Contract Without Quantities, 1980 Edition (Private or Local Authorities version as appropriate) is commonly used.

D. Based on a schedule of rates (sometimes known as *measured contracts* or *measurement contracts*)

This method operates in a similar way to that described in B. above, tenders being based upon a schedule of rates as explained in (i)–(iii) below. A particular advantage arising from its use is that it allows for a contract to be signed and work to start on site when the design is only in outline form, and in consequence the pre-contract period is reduced considerably.

(i) Standard schedule

A standard schedule lists under appropriate trade headings all the items likely to arise in any construction project, with a unit rate against each item. The best-known of such schedules is the *Schedule of Rates for Building Works* prepared by the Property Services Agency of the Department of the Environment.[2]

Tenderers are asked to tender percentage additions (or deductions) to the listed rates, usually by sections or sub-sections, thus allowing for variations in construction costs since the date of the preparation of the schedule used.

Advantage

1 Tenderers using a particular schedule often soon become familiar both with the item descriptions and the rates and are able to assess percentage adjustments relatively easily.

Disadvantages

1 In comparing and assessing a range of tenders, the surveyor has the task of gauging the effect of a series of variables, making the choice of the most favourable tender difficult.
2 The parties are unable to have a precise indication of their respective commitments.

(ii) 'Ad hoc' schedule

This is a schedule specially prepared for a particular project and lists only those items which are appropriate to that project, including any special or unusual items. An 'ad hoc' schedule may be pre-priced by the surveyor (in which case the form of the tender will be the same as when using a standard schedule) or the rate column may be left blank by the surveyor for the tenderer to insert individual rates against each item. The latter method, because of the absence of quantities, makes the comparison and assessment of tenders much more difficult.

Advantages

1   Tenderers are only required to concern themselves with a restricted range of items, thus enabling them to assess rates or percentages more accurately.
2   Tenderers are able to obtain a clearer picture of the scope of the work from the items listed in the schedule.

Disadvantages

These are similar to those applying to standard schedules.

(iii) Bills of quantities from previous contract

The bill of quantities used will normally be for a comparable type of building of similar constructional form to the proposed project. It is, in effect, a pre-priced 'ad hoc' schedule and will be used in the same way.

   This is the method of tendering normally used in *serial tendering*, described on p. 49.

Advantages

1   The time required to prepare tender documents is reduced to the minimum.
2   Tenderers have to consider only a restricted range of items.

Disadvantages

1   The parties are unable to have a precise indication of their respective commitments.
2   There may be a considerable discrepancy between the successful tender and the real cost of the work, owing to the approximate nature of the quantities.

E. Based on cost reimbursements (also known as *prime cost* or *cost plus* because the method of payment is by reimbursement to the contractor of his prime cost, plus a management fee)

There are three variants of this type of contract, distinguished by the way in which the fee is calculated. Each is dealt with separately below.*

'Prime cost' means the total cost to the contractor of buying materials, goods and components, of using or hiring plant and of employing labour, in order to carry out construction works.

Of all the types of contract, this produces the most uncertainty as to the financial outcome. Tenders contain no total sum and it may be very difficult to form any reliable estimate of the final cost.

It is widely recognized as the most uneconomical type of contract and therefore is one which normally should be used only in circumstances where none of the other types is appropriate. Because of the possibility of inefficiency and waste of resources, contracts of this type need to embody provisions giving the architect some control over the level of labour and plant employed.

One of the attractions of prime cost contracts is that work on site can commence in the early stages of design and this may be all-important to the client. There may be circumstances where, to the client, cost is a less important factor than time. Consequently, a start on site at the earliest possible time may be financially more advantageous in the long run than a lower final cost of construction which might have resulted from the use of another type of contract.

It should be noted that no site measuring is necessary other than as checks on the quantities of materials for which the contractor submits invoices, and for purposes of checking nominated sub-contractors' and nominated suppliers' accounts.

The process of calculating and verifying the total prime cost involves a vast amount of investigation and checking of invoices, time sheets, sub-contractors' accounts, etc., which can be both tedious and time-consuming. It is therefore in the interests of both the surveyor and the contractor that at the outset a proper

* The JCT Standard Form of Prime Cost Contract, 1992, is commonly used.

system of recording, verifying and valuing the prime cost is agreed and strictly implemented. In addition, it is vitally important to define clearly what is intended to be included as prime cost and what is intended to be covered by the fee. The standard form for the fixed fee variant of this type of contract does so in the appended schedules.[3] The publication *Definition of prime cost of daywork*[4] may be adapted for this purpose if the standard form is not used.

The advantages and disadvantages of prime cost contracts over lump sum and schedule of rates contracts follow.

Advantages

1 The time required for preparation of tender documents and for obtaining tenders is minimized, thus enabling an early start on site to be made.
2 Work on site may proceed before the detailed design is complete.

Disadvantages

1 The parties have the least precise indication of their respective commitments.
2 The cost of construction to the client is likely to be greater than if other types of contract were to be used.
3 The computation and verification of the total prime cost is a long and tedious process.

The variants below differ in the way in which the fee for the contractor's services is determined. Variants (ii) and (iii) are the consequence of a general acknowledgement that it is desirable to provide an incentive to economize in the use of resources on the part of the contractor.

(i) Cost plus percentage fee

The contractor is paid a fee equal to an agreed percentage of the prime costs of labour, materials and plant used in carrying out the work.

The outstanding disadvantage (to the client) is that the more inefficient the contractor's operations are and the greater the waste of resources, the higher the fee paid to the contractor will be. (To counter this, the percentage is sometimes made to vary inversely as the prime cost.)

The following is a simple illustration of this contractual arrangement, which at the same time will provide a basis for comparison with the fixed fee form. A contract is assumed where the *estimated* total cost of materials, labour, and plant was £50,000; the contractor's overheads were calculated by him as 15 per cent and he required 5 per cent for profit. He therefore tendered at 20 per cent overall addition to the prime cost as his fee and his tender was accepted. Assuming that the estimate of prime cost proved at the completion of the contract to be accurate, the total final cost to the client would be:

	£	£
Total prime cost		50 000
15% addition for overheads	7500 ⎱ 20%	
5% addition for profit	2500 ⎰	10 000
Total cost of contract		60 000

If, however, due to uneconomic organization of the contract, inefficiency and excessive waste, the total prime cost was £55,000 for the same job, then the total cost would be:

	£
Total prime cost	55 000
20% addition for overheads and profit	11 000
Total cost of contract	66 000

As the contractor's overheads chargeable to this job would still be £7500, it follows that the real profit would be £3500. The disincentive to the contractor to work efficiently is thus seen to be strong.

(ii) Cost plus fixed fee

Under this variant, the fee paid to the contractor is a fixed sum which normally does not vary with the total prime cost, but is based on an estimate of the likely total. The only ground on which the fee might be varied is if either the scope of the work or the conditions of carrying it out were to be materially altered after the contractor tendered.

It should be noted that the fee, if considered in percentage terms, is lower when the prime cost is higher.

Using the same illustration as before, but with the contractor having tendered a fixed fee of £10,000 (made up of £7500 for overheads and £2500 for profit), the sum paid to the contractor as fee would be equivalent to 20 per cent if the total prime cost was £50,000.

If as before, the prime cost was higher for the same reasons, the financial picture would be:

	£	£
Total prime cost		55 000
Overheads	7500	
Profit	2500	10 000
Total cost of contract		65 000

The fee is now equal to 18.18 per cent of the prime cost and the profit portion only 4.55 per cent. If the prime cost rose to £60,000, the fee would equal only 16.67 per cent and the profit portion 4.17 per cent.

(iii) Target cost

This variant is really one or other of the two preceding ones with another factor added. As an incentive to reducing the total prime cost, the agreement provides for a bonus to be paid to the contractor if the total cost is less than an agreed sum (the 'target') and also a penalty to be paid by him if the total cost exceeds that sum. The bonus and penalty are commonly 50 per cent of the difference between the total amounts, but may be any agreed percentages. The target cost is an estimate of the likely total cost.

Taking the same illustration again, it is assumed that the fixed fee method of payment is to be used, and that the agreed sum based on an estimate of likely total cost is £60,000. At the end of the job (assuming the higher cost shown above), the final payment would be calculated like this:

	£
Total prime cost	55 000
Fixed fee	10 000
	65 000
Deduct penalty, being 50% of excess over £60 000	2 500
Amount of final payment	62 500

In effect, the fee has been reduced to £7500, which is only just enough to meet overheads, leaving nothing for profit. It follows therefore that the contractor will need to be very satisfied that the 'agreed sum' is a realistic estimate of likely cost and the surveyor must give him every reasonable opportunity and assistance to satisfy himself in that regard.

However, if the contractor were able to reduce the prime cost, say to £48,000, final payment would be calculated as:

	£
Total prime cost	48 000
Fixed fee	10 000
	58 000
Add bonus, being 50% of saving over £60 000	1 000
Amount of final payment	59 000

The fee has now increased to £11,000, including £3500 for profit, the overheads remaining at £7500. As a percentage, the profit is 7.29 per cent of the prime cost. Thus, under this contractual arrangement, an extra cost or saving is shared between the client and the contractor.

The composite nature of contracts

Although, for convenience, classifications and type labels such as the foregoing are commonly used, in practice contracts often combine the characteristics of two or more types. McCanlis has pointed out[5] that, for example, a lump sum contract based on bills of firm quantities often contains items with provisional quantities requiring remeasurement and therefore such items bear the characteristics of a schedule of rates contract. Also, the provisional sums included in the bills for daywork and expended on work which is not readily measurable or not reasonably priceable as measured work, form a prime cost plus percentage fee contract within the main lump sum contract. Thus, many contracts which are regarded as of the lump sum type are, in reality, a combination of several types. Other examples of the composite nature of many contracts are given in the reference above.

Circumstances in which the various types of contract may be used

Usually the circumstances peculiar to a project will indicate which type of contract is most appropriate. Occasionally, more than one type might be suitable, in which case, the one which seems to offer the greatest benefits to the client should be chosen as the client is the one who will be paying the bill.

The suitability of the types of contract discussed earlier may be related to varying circumstances as follows.

Based on firm bills of quantities

(i)   When there is time to prepare a sufficiently complete design to enable accurate quantities to be measured.

(ii) When the client's total commitment must be known beforehand, for example, in order to make adequate borrowing arrangements or when approval to the proposed expenditure has to be obtained from a finance or housing committee of a local authority, from a central government department or from a board of directors.

Based on bills of approximate quantities

(i)   When the design is fairly well advanced but there is insufficient time to take off accurate quantities or the design will not be sufficiently complete soon enough for that to be done.

(ii) When it is desired to have the advantages of detailed bills but without the cost in terms of time and/or money.

Based on drawings and specification

(i) When the project is fairly small, i.e. up to a value of, say, £50,000.

(ii) Where time is short and the client considers it to be less important to have the benefits of bills of quantities than the early completion of the construction work, while retaining the advantages of a lump sum contract.

Based on a schedule of rates

(i) When the details of the design have not yet been worked out or there is considerable uncertainty in regard to them.

(ii) When time is pressing.

(iii) When *term contracts* are envisaged. These are appropriate where a limited range of repetitive work is required to be carried out, such as the external redecoration of an estate of houses. The contractor tenders on the basis of unit rates, which are to remain current for the stated 'term', usually one year or the estimated period which the proposed work will take.

Based on prime cost plus fee

(i) When time is short and cost is not as important as time.

(ii) When the client wishes to use a contractor who has worked satisfactorily for him before and whom he can trust to operate efficiently, while being prepared to pay the higher cost entailed in return for the advantages.

(iii) In cases of emergency, such as repairs to dangerous structures.

(iv) For maintenance contracts.

(v) For alterations jobs where there is insufficient time or it is impracticable to produce the necessary documentation.

The following examples will illustrate some of the foregoing considerations.

Example 1

A client company owned a department store in a prime position in Manchester. It had acquired the adjoining site for an extension which it was estimated would cost £500,000. The Board of Directors wanted the work to be done by a particular contractor who had done work for them before.

Apart from the choice of the contractor, the important consideration from the client's point of view was the need to commence trading in the new extension at the earliest possible date and also to restrict the disturbance to their normal trading operations to as short a period as possible.

Of the variants to the lump sum type of contract, only the drawings and specification type could be considered on the ground of time but because of the early start requirement it was rejected. As it was thought unlikely that there would be many variations during the course of construction, the schedule of rates type was rejected too.

The clients were not so concerned about cost as about time, and because they knew the contractors and were satisfied that they would do an efficient and speedy job, the type of contract eventually recommended (and used) was the target cost form of prime cost plus a fixed fee. The contractors produced an estimate of cost of £530,000 and after some negotiation the target cost was finally agreed at £518,000.

Example 2

The clients were manufacturers of clothing and had a new factory in the early stages of construction in a suburb of Birmingham. The contract was based on bills of quantities and the work was proceeding on schedule. Another site had been acquired by the clients for a building of similar size and form of construction about 18 miles away at Coventry. The clients were anxious to begin production as soon as possible in both factories.

As the two sites were reasonably close and time was an important factor, the type of contract based on a pre-priced schedule of rates (namely, the priced bills of quantities for the first contract) seemed the best choice. The prices in those bills could only be used, of course, if the same contractor were able and willing to take on the second contract as he held the copyright in the prices. In the event, he was prepared to do so and percentage additions to each of the work sections in the bills were negotiated between the surveyor and the contractor's estimator. A new 'Preliminaries' section of the bills appropriate to the second project was prepared and prices negotiated for the items in it.

Example 3

An Inner London Borough Council was proposing to improve some 1500 of its older dwellings by replacing all the sanitary appliances with modern equipment over the next three or four years. Finance had already been allocated for the work to commence in the current financial year and a start was to be made as soon as contracts could be arranged.

The main characteristic of this situation was the large amount of repetition of relatively few items of work. Time was also a factor in so far that an early start was essential if allocated finance during the current financial year was to be fully used.

The circumstances all pointed to 'term contracts', and, accordingly, contracts based on *ad hoc* schedules were used. Schedules of the plumbing and associated builder's work items were drawn up for tenderers to insert their own rates against the items. Provision was made in the contracts for the rates to be updated at 12-monthly intervals.

The risk factor

An important element of construction contracting is risk, i.e. the risk the parties take that the implementation of the contract may be detrimental in some degree to their financial or other interests. On his part, the contractor takes the risk that his anticipated profit will be reduced or converted into a loss as a result of the outworking of the conditions under which the contract is carried out. On his part, the Employer takes the risk that he will become liable for a greater total cost than he envisaged when initiating the project.

The types of contract described above carry a varying degree of risk for the parties. Generally speaking, the contractor's risk increases as the Employer's risk reduces and *vice versa*. Thus, the contractor bears a high degree of risk where the contract is based on drawings and specification only, for two reasons. First, as it is a lump sum contract, he must estimate his expected costs as accurately as possible because any adverse mistake will reduce his profit. Secondly, there being no bill of quantities provided for tendering purposes, he must take off his own quantities from the drawings in order to formulate his tender, and, again, any error he may make in the process will affect his profit margin.

The Employer's risk is small in the situation just described. He knows at the outset what his financial liability will be and is under no contractual obligation to reimburse the contractor for any errors which he may have made in preparing his tender.

At the other end of the spectrum, where a cost reimbursement contract is used, the contractor's risk is reduced considerably because he is paid his full costs and a fee in addition. His only risk is in pitching at the right level the fee which he tenders but, even so, he is most unlikely to make a loss. On his part, however, the Employer bears the risk of the prime cost becoming much higher than estimated, owing perhaps to an inefficient site agent or to wastage of resources.

Other types of contract fall within these two extremes and Figure 2.1 indicates a ranking of them according to the allocation of risk borne by the parties.

Relationships between the parties

In all the traditional types of contract, the pattern of relationships between all the parties is normally similar. Figure 2.2 illustrates these relationships and indicates the lines of communication between them.

Contractual links exist:

(i)   between the Employer and each of his professional advisers (architect, quantity surveyor and consultant engineers);
(ii) between Employer and contractor;
(iii) between the contractor and each sub-contractor.

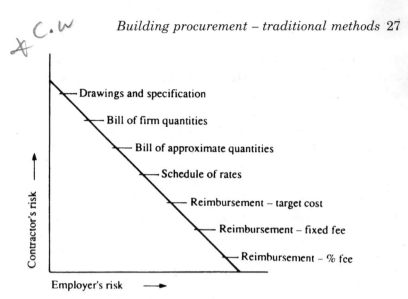

Figure 2.1 Allocation of risk

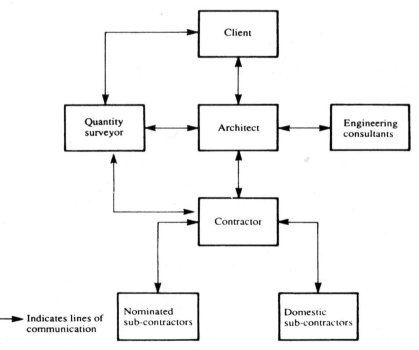

Figure 2.2 Relationships and communication links

There may also be Warranty Agreements between the Employer and sub-contractor if the Employer wishes (as, for example, by use of Form NSC/W in conjunction with JCT 80).

References

1. RICS, Contracts in Use, *Chartered Quantity Surveyor*, 1989, Vol. 11, No. 5, p. 24.
2. *Schedule of Rates for Building Works, 1985* (London: HMSO, 1985).
3. *Standard Form of Prime Cost Contract*, 1992 (London: RIBA Publications Ltd, 1992).
4. *Definition of prime cost of daywork carried out under a building contract* (London: RICS Books, 2nd edn, 1975).
5. McCANLIS, E. W., *Tendering Procedures and Contractual Arrangements* (London: RICS Books, 1967), p. 10.

Bibliography

1. CLAMP, H. and COX, S., *Which Contract?* (London: RIBA Publications, 1989).
2. CHAPPELL, David, *Which Form of Building Contract?* (London: ADT Press, 1991).
3. TURNER, Dennis, *Building Contracts: A Practical Guide* (London: Longman, 5th edn, 1994).

3

Building procurement – alternative methods

Since the early 1960s, various alternative ways of promoting and carrying out construction projects have been devised with varying degrees of success. The reasons why many building owners and developers considered that the traditional procedures were no longer satisfactory were:

(i) The rapidly spiralling cost of construction meant that large sums of money had to be borrowed to finance projects;

(ii) high interest rates meant that the time occupied by the traditional procedures resulted in substantial additions to the construction cost;

(iii) clients were becoming more knowledgeable on construction matters and were demanding better value for money and an earlier return on their investment;

(iv) high technology installations required a higher quality of construction.

Attention has focused on reducing the time traditionally occupied in producing a design and preparing tender documentation, thus enabling construction work to begin sooner.

Another important factor has been bringing the contractor in at an early stage in the design of a project. Under the traditional procedures, the contractor rarely played any part until the tender stage was reached, after virtual completion of the design. The increasing complexity of projects, however, led to the realization that it was in the interests of clients and architects to use the vast amount of knowledge and practical experience of contractors early in the design process, and that this would make a valuable contribution to a successful outcome.

The following are the principal alternative methods of building procurement now in use.

A. Design and build (sometimes called *design and construct* or *package deals*)

When first introduced in its modern form, this method usually meant the contractor using an industrialized building system which was adapted to meet the client's requirements – usually straightforward, rectangular warehouse or factory buildings with a minimum of design. More recently, the method has been applied to a wide range of building types.

Essentially the contractor is responsible for the design, for the planning, organization and control of the construction and for generally satisfying the client's requirements, and offers his service for an inclusive sum. A proprietary, prefabricated building system may or may not be used.

The procedure is initiated by the client (or an architect on his behalf) preparing his requirements in as much or as little detail as he thinks fit. These are then sent to a selection of suitable contractors, each of whom prepares his proposals on design, time and cost, which he submits together with an analysis of his tender sum. The client then accepts the proposals he is satisfied best meet his requirements and enters into a contract* with the successful tenderer. The latter then proceeds to develop his design proposals and to carry out and complete the works.

The client may use the services of an independent architect and quantity surveyor to advise him on the contractor's proposals as to design and construction methods and as to the financial aspects respectively. He may also appoint an agent to supervise the works and generally to act on his behalf to ensure that the contractor's proposals are complied with.

Figure 3.1 shows the normal pattern of relationships and indicates the lines of communication between the parties in a design and build contract.

---

* The JCT Standard Form of Building Contract With Contractor's Design 1981 Edition is a suitable form of contract.

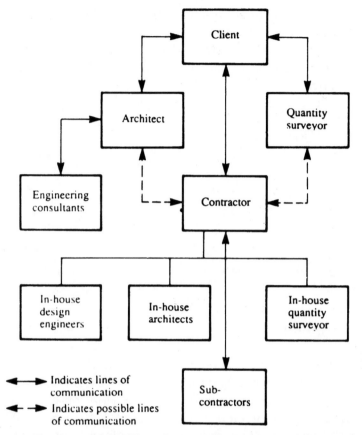

Figure 3.1 Design and build contracts: relationships of the parties

Contractual links exist:

(i)   between the client and the contractor;
(ii) between the contractor and each sub-contractor;
(iii) between the client and each of his independent advisers.

Advantages

1   Single point responsibility is provided, i.e. the contractor is solely responsible for failure in the design and/or the construction.
2   The client has only one person to deal with, namely, the contractor, whose design team includes architects, quantity surveyors, structural engineers, etc.

3   The client is aware of his total financial commitment from the outset.
4   Close intercommunication between the contractor's design and construction teams promotes co-operation in achieving smoother running of the contract and prompt resolution of site problems.

Disadvantages

1   Variations from the original design are discouraged by the contractor and, if allowed, are expensive.
2   The client has no means of knowing whether he is getting value for money unless he employs his own independent advisers, which adds to his costs.
3   If the contractor's organization is relatively small, he is unlikely to be as expert on design as he is on construction, and the resulting building may be aesthetically less acceptable.

B. Management contracting

This method has been increasing in popularity since the early 1970s. A study of management contracting carried out in 1985 by the Centre for Construction Market Information estimated that the output of management contracting in 1984 was £890m, with a growth in 1985 of 9 per cent. A survey of usage of various forms and types of contract carried out in 1988[1] showed that management contracts were used on 9.41 per cent of projects (by total value) in 1987 compared with 14.4 per cent in 1985 and 12.01 per cent in 1984.

The principal characteristic of management contracting is that the management contractor does none of the construction work himself but it is divided up into work packages which are sub-let to sub-contractors, each of whom enters into a contract with the management contractor. The latter is normally either nominated by the client on the basis of the contractor's previous experience of management contracting or is selected by competition based upon tenders obtained from a number of suitable contractors for (a) the management fees and (b) prices for any additional services to be provided before or during the construction period (unless to be paid for on a prime cost basis, in which case the tenders will include percentage additions required to the respective categories

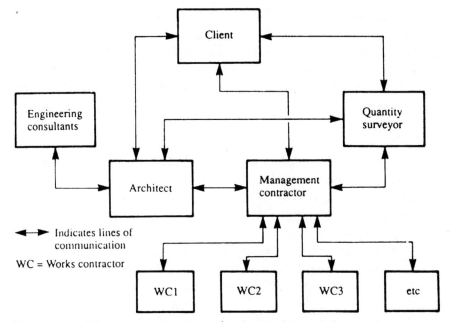

Figure 3.2 Management contracts: relationships of the parties

of prime cost). The successful contractor will then enter into a contract* with the client.

The management contractor's role therefore is that of providing a construction management service on a fee basis as part of the client's management team – organizing, co-ordinating, supervising and managing the construction works in co-operation with the client's other professional consultants. As part of his service, he provides and maintains all the necessary site facilities, such as offices, storage and mess huts, power supplies and other site services, common construction plant, welfare, essential attendances on the works contractors (e.g. unloading and storing materials, providing temporary roads and hardstandings and removing rubbish and debris) and dealing with labour relations matters.

Figure 3.2 shows the usual pattern of relationships and indicates the lines of communication between the parties to a management contract.

* The JCT Standard Form of Management Contract 1987 Edition is a suitable form of contract.

Contractual links exist:

(i)   between the client and each member of the design and management team (including the management contractor);

(ii) between the management contractor and each of the sub-contractors.

Advantages

1   Work can begin on site as soon as the first one or two works packages have been designed.

2   Overlapping of design and construction can significantly reduce the time requirement, resulting in an earlier return on the client's investment.

3   The contractor's practical knowledge and management expertise are available to assist the design team.

4   Where the nature and extent of the work may be uncertain, as in refurbishment contracts, the design of later work packages may be delayed until more information becomes available as the work progresses, without extending the construction period.

5   The contractor, being part of the client's team, is able to identify with the client's needs and interests.

6   Because works contracts are entered into close to the time of their commencement on site, they can be based on firm price tenders.

Disadvantages

1   Uncertainty as to the final cost of the project until the last works contract has been signed.

2   The number of variations and the amount of remeasurement required may be greater than on traditional contracts because of the greater opportunity to make changes in design during the construction period, because of problems connected with the interface between packages, and because packages are sometimes let on less than complete design information.

C. British Property Federation (BPF) system

This system was introduced at the end of 1983 by the British Property Federation, which represents property, development, insurance and financial organizations who are the customers of the construction industry. It published a *Manual* at that time describing the system in detail, and in 1984 published the BPF Edition of the ACA Form of Building Agreement.*

The System claims 'to produce good buildings more quickly and at lower cost' and 'to remove as much as possible of the overlap of effort between designers, quantity surveyors and contractors'. It operates by a progression through five stages, as follows:

Stage 1 – *Concept*: during which the client produces an outline brief and appoints a client's representative.

Stage 2 – *Preparation of the brief*: during which a design leader is appointed to prepare a design programme and the client's representative develops the brief containing a cost plan, time limits and basic design requirements; he also sets a target cost.

Stage 3 – *Design development*: during which the design leader expands the brief into drawings and a specification.

Stage 4 – *Tender documentation and tendering*: during which tender documents (drawings, a time schedule and a schedule of activities or bills of quantities) are produced and competitive tenders obtained.

Stage 5 – *Construction*: during which the client's representative has responsibility for all aspects of management, a supervisor monitors the works on behalf of the client, and a site manager is the full-time representative of the contractor on site.

The BPF System therefore broadly follows the traditional procurement pattern but is different in the following respects:

(i) the client's representative, who is named in the contract, is in control in the role of project manager, no architect, as such, being named in the contract;
(ii) the client's representative (unlike the architect under JCT contracts) may delegate any or all of his duties to any number of persons or firms;

* The JCT Standard Form of Building Contract With Contractor's Design 1981 Edition is a suitable alternative form of contract.

(iii)   a priced schedule of activities may replace the traditional bill of quantities, with the intention of reducing the length of time (and therefore cost to the client) necessary for preparing contract documents;

(iv)   a proportion of the design work is transferred to the contractor, the amount depending upon the extent of design included in the tender documents;

(v)   variations, it is claimed, are reduced to a minimum, by all concerned being encouraged to make prompt decisions;

(vi)   there is no provision for nominated sub-contractors or nominated suppliers but certain work or certain goods may be stipulated as to be carried out or to be supplied by a person or one of a list of persons named in the contract documents;

(vii)   the design leader and other consultants are paid a lump sum fee instead of a percentage of the value of the project;

(viii)   disputes are settled by a named adjudicator in order to speed up settlement, with arbitration in reserve if the parties are not satisfied with his decisions and for those matters not specified in the contract as referable to the adjudicator.

Figure 3.3 shows the pattern of relationships between the parties in a contract using the BPF System and indicates the lines of communication between them.

Contractual links exist:

(i)   between the client and the client's representative;
(ii)   between the client and the design leader;
(iii) between the client and each design consultant;
(iv)   between the client and the supervisor;
(v)   between the client and the contractor.

Advantages

1   Some or most of the design work may be transferred to the contractor, thereby obtaining the benefits of his expertise and also shortening the pre-contract period.
2   All those working on a project are aware of their respective duties and responsibilities, as they are clearly defined in detail in the BPF Manual.
3   The matters to be specified or shown in the brief, the master cost plan and in the tender documents are fully detailed in the Appendixes to the Manual.

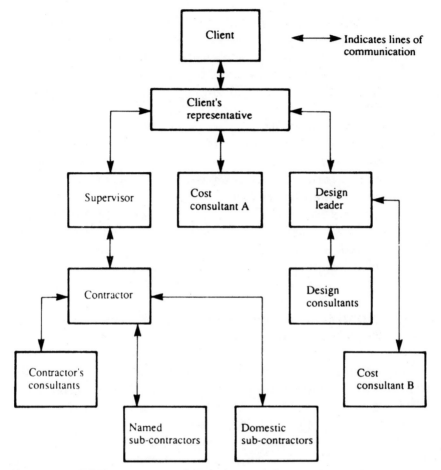

Figure 3.3 BPF contracts: relationships of the parties

4 The prescribed management structure is calculated to induce a greater sense of urgency in the minds of the consultants, and the system is designed to reduce as far as possible the total project period.

Disadvantages

1 Where a bill of quantities is not provided by the client, the scope of the work covered by the contractor's price is less precisely defined.
2 If a bill of quantities is not provided, errors of omission by a successful tenderer inevitably lead to claims and/or to cutting corners in an attempt to recover the consequential loss.

3   The pricing of variations will lead to more protracted negotiations between the client's representative and the contractor where a bill of quantities is not used.

4   Some contractors may not be sufficiently competent in interpreting and developing designs from conceptual information.

5   The use of one contractor in the design process inevitably brings in also those sub-contractors with whom that contractor normally works, which may leave the client in the position of feeling obliged to use those sub-contractors subsequently without competition.

6   The dominant role of the client's representative could lead to contractors being somewhat less than satisfied that they will be treated with the same impartiality as they have come to expect from independent architects and quantity surveyors.

D. Construction management (sometimes called the *separate contracts system*)

Construction management is not a procurement system in itself but is only part of the total process. It is a professional consultant service to the client, provided on a fee basis, with the design and construction services being provided by other organizations.

The construction manager is responsible for the organization and planning of the construction work on site and for arranging for it to be carried out in the most efficient manner. The construction work itself is normally carried out by a number of contractors, each of whom is responsible for a defined work package. All the work packages together constitute the total project. Each works contractor enters into a direct contract with the client.

The construction manager's duties normally include any or all of the following:

(i)   co-operation and consultation with the other members of the client's professional team;

(ii)   preparation and updating of a detailed construction programme;

(iii) preparation of materials and components flows and arranging for advance ordering;

(iv)   determining what site facilities and services are required and their location;

(v)   breaking down the project into suitable work packages in consultation with the other members of the client's team and recommending suitable contractors to be invited to tender for work packages;

(vi)   obtaining tenders from contractors and suppliers;

(vii) evaluating tenders and making recommendations on them to the client's team;

(viii) co-ordinating the work of the works contractors to ensure that it is carried out in accordance with the master programme;

(ix)   establishing all necessary management personnel on site with responsibility to manage and supervise the project;

(x)   dealing with any necessary variations to the work, providing the design manager with estimates of their likely cost and subsequently issuing instructions to works contractors;

(xi)   submitting to the quantity surveyor applications from works contractors for periodic payments and all necessary documentation enabling the final accounts of works contractors to be settled.

Figure 3.4 shows the normal pattern of relationships between the parties and indicates the lines of communication between them.
   Contractual links exist:

(i)   between the client and the construction manager;

(ii)   between the client and each of the works contractors;

(iii) between the client and each of his professional advisers (architect, quantity surveyor, engineering consultants, etc.).

Advantages

1 The construction work is more closely integrated into the management of the project.

2 Close liaison between the construction manager and design manager leads to prompt identification of and decisions relating to practical problems.

3 Detailed design can continue in parallel with construction, work packages being let in succession as the design of each is completed, thus shortening the project time.

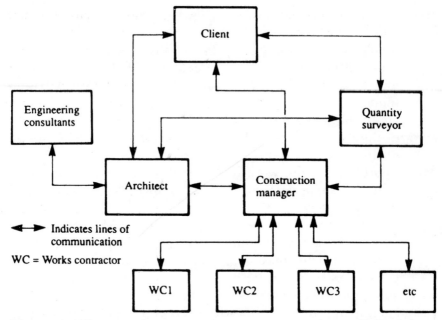

Figure 3.4 Construction management contracts: relationships of the parties. *Note:* Lines of communication also exist between the architect and surveyor and each of the works contractors. To avoid confusion, these lines are omitted.

4   Privity of contract between the client and each of the works contractors provides the client with a readier means of redress in the event of difficulties, such as delays, arising.

Disadvantages

1   The client has one more consultant and a number of contractors with whom to deal instead of only one main contractor.
2   The client's financial commitment is uncertain until the last of the works contracts has been signed.

E. Project management

Project management is not a procurement system in itself, in that it does not include the site construction process but only its general supervision. It has been defined[2] as 'the overall planning, control and co-ordination of a project from inception to completion aimed at meeting a client's requirements and ensuring completion on time, within cost and to required quality standards'.

A number of project management organizations came into existence during the late 1970s and 1980s to meet the need for the management of projects which were becoming larger and increasingly complex. Many clients do not have the in-house skills and experience necessary for the successful management of construction projects and so need to employ an independent project management company to do it for them.

Quantity surveyors, by their training and experience in financial and contractual matters, coupled with a detailed knowledge of construction processes, are well qualified to offer a project management service, and a number of established quantity surveying practices have set up associated companies offering such a service. Other groups of professionals, such as architects, engineers and building and valuation surveyors, are also now filling this role.

The project manager, in effect, becomes the client's representative, with authority to supervise and control the entire planning and building operation from acquisition of the site to completion of the project and settlement of the accounts. The service he provides is essentially one of planning, organizing and co-ordinating the services provided by surveyors and lawyers in relation to site acquisition; the architect, engineers and quantity surveyor in relation to project planning and design; and the contractor and sub-contractors in carrying out the site construction work; but does not include the carrying out of any of their duties himself.

Figure 3.5 shows the normal pattern of relationships between the parties and indicates the lines of communication between them.

Contractual links exist:

(i)   between client and project manager;
(ii)   between client and each of his professional advisers;
(iii) between client and contractor;
(iv)   between contractor and each sub-contractor.

F. Joint venture

This is an arrangement for building procurement which has developed out of the increasing complexity of construction projects,

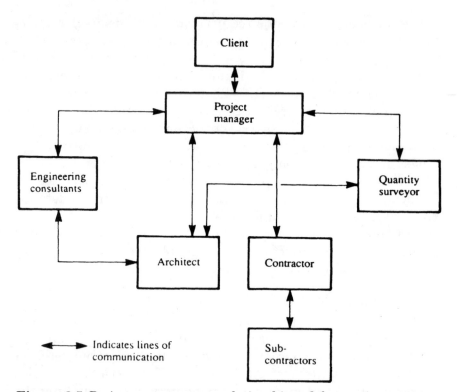

Figure 3.5 Project management: relationships of the parties

often with unorthodox methods or sequences of construction. It is applicable to large-scale projects of, say, £5m value or more, where there is a higher than normal proportion of engineering and other specialist services. It has been defined as follows:

> A partnership between two or more companies covering building, mechanical and electrical engineering, or other specialist services for the purpose of tendering for, and executing a building or civil engineering contract, each of the participating companies having joint and several liability for their contractual obligations to the employer.[3]

The basis of such projects is a single contract to which all the partner companies to the joint venture are signatories. Any of the standard forms of contract in current use may form the contract document, and in the publication just quoted the NJCC recommends the use of JCT 80 for projects not primarily of an engineering nature. Modifications will need to be made to cover joint and several liability of the participating companies and to provide for any performance warranty which the Employer may require.

Joint venture has the effect of making specialist contractors, who under more orthodox arrangements would be sub-contractors, partners with the building contractor. This does not preclude the use of sub-contractors (whether nominated, named or listed), a role which is filled by secondary specialist contractors who are not partners in the joint venture.

Advantages

1 The combined resources of participating companies result in greater economy in the use of specialist manpower, design and equipment.
2 There is a greater degree of unified action between the contracting parties.
3 Lines of communication between the Employer's professional advisers and the building contractor and specialist contractors are shorter and better-defined.
4 Improved integration of work sequences results in a shorter contract time and fewer management problems.

Disadvantages

1 If one JV partner withdraws, the remaining partner or partners must accept total liability to complete the project.
2 A separate, unified JV bank account must be arranged and maintained by all the partners.
3 Separate insurance policies must be taken out on behalf of all the JV partners, specifically for the proposed project.
4 A board of management with a single managing director needs to be appointed jointly by the JV partners to ensure unified actions.

The pattern of relationships in a Joint Venture project and the lines of communication between the parties are similar to those illustrated in Figure 2.2, but with the JV partners together fulfilling the role of 'Contractor'.

Contractual links exist:

(i)   between the JV participating companies;
(ii)   between the Employer and the JV partners collectively;
(iii)   between the Employer and each of his professional advisers;
(iv)   between the JV partners and each of the sub-contractors.

References

1. RICS, Contracts in Use, *Chartered Quantity Surveyor*, 1989, Vol. 11, No. 5, p. 24.
2. RICS, Project Management: Defining the Terms, *Chartered Quantity Surveyor* 1986, Vol. 8, No. 11, p. 20.
3. NJCC, *Guidance Note 1 – Joint Venture Tendering for Contracts in the UK* (London: RIBA Publications Ltd, 1985), p. 1.

4

Tendering methods and procedures

Methods of contractor selection

The selection of a contractor to carry out a construction project is an important matter requiring careful thought. A wrong choice may lead to an unhappy client/contractor relationship, a dissatisfied client and possibly even an insolvent contractor.

Any one contractor will not be suitable for any one job. A contracting organization will be geared to work of a particular size or price range and will be unsuitable or uneconomic for contracts outside that range. For example, a national company would probably not be suitable to carry out a small factory extension or to build a single house on an infill site in an urban location. The overhead costs of that size of organization would be such as to make it impossible for the company to carry out the work for an economic price.

At the other end of the scale, a small firm, used only to jobbing work and small alterations and extensions, would probably be out of its depth in handling a contract for the erection of even a modest-sized reinforced concrete office block. Such a firm, with little experience of estimating and probably none in the pricing of bills of quantities, would be likely to tender at too low a level by failing to include for all the costs involved. If given the contract, the firm would no doubt soon get into difficulties and probably would end up in disaster.

The client's professional advisers should aim to find a contracting company (a) that is financially stable and has a good business record, (b) for which the size of the project is neither too small nor too large, (c) that has a reputation for good-quality workmanship and efficient organization and (*d*) that has a good record of industrial relations.

There are three principal methods of choosing a contractor: (*a*) open tendering, (*b*) selective tendering and (*c*) nomination.

Open tendering

This is initiated by the client's project manager, architect or quantity surveyor advertising in local newspapers and/or the technical press, inviting contractors to apply for tender documents and to tender in competition for carrying out the work, the main characteristics of which are given. Usually a deposit is required in order to discourage frivolous applications, the deposit being returnable on the submission of a *bona fide* tender.

Advantages

1   There can be no charge of favouritism as might be brought where a selected list is drawn up (this is of concern particularly to local authorities who, probably for this reason more than any other, tend to use open tendering more than other clients).
2   An opportunity is provided for a capable firm to submit a tender, which might not be included on a selected list.
3   It should secure maximum benefit from competition (it may not always do so, however, as may be seen from the disadvantages below).

Disadvantages

1   There is a danger that the lowest tender may be submitted by a firm inexperienced in preparing tenders (particularly if bills of quantities are used) and whose tender is only lowest as a consequence of having made the most or the largest errors.
2   There is no guarantee that the lowest tenderer is sufficiently capable or financially stable. Although obtaining references will provide some safeguard, there may be little time in which to do so.
3   Total cost of tendering is increased as all the tenderers will have to recoup their costs eventually through those tenders which are successful. The result can only be an increase in the general level of construction costs.

Open tendering was deprecated in the Banwell Report.[1] Consequently central government departments and most local authorities do not use this method of obtaining tenders. There tends, however, to be an increase in its use by local authorities in times of national economic difficulties with attendant public expenditure cuts, presumably because it is assumed that the more tenderers there are, the lower the price for which the work can be carried out.

Selective tendering

Selective tendering may be either *single stage* or *two stage*, depending on whether the full benefits of competition are desired (in which case single stage tendering is used) or whether limited competition plus earlier commencement of the works on site is considered advantageous (using two stage tendering). The NJCC has issued a *Code of Procedure* for each method.[2] It has also issued a Code for Design and Build contracts, using either the single stage or the two stage procedure,[3] and recommendations for tendering for industrialized building projects.[4] (Further references to 'the *Codes*' in this chapter relate to the first two.)

Single stage selective tendering

Under this procedure, a short list is drawn up of contractors who are considered to be suitable to carry out the proposed project. The names may be selected from an approved list or 'panel' maintained by the client (as many public authorities do), or may be specially chosen.

In the latter case the contractors may be invited, through suitably worded advertisements in the press, to apply to be considered for inclusion in the tender list. This gives the client the opportunity to exclude any firms thought to be unsuitable and to limit the number of tenderers. At the same time, it gives any firm the opportunity to apply to be considered.

It is recommended that the number of tenderers should be limited to between five and eight, depending on the size and nature of the contract. If the firms on the list are all ones which are reputable, well-established and suitable for the proposed work, and the client fixes the construction time, then the selection is resolved into a question of price alone and the contract can be safely awarded to the firm submitting the most favourable tender.

Advantages

1   It ensures that only capable and approved firms submit tenders.
2   It tends to reduce the aggregate cost of tendering.

Disadvantage

1   The cost level of the tenders received will be higher, owing to there being less competition and also to the higher calibre of the tenderers.

Nomination

This is sometimes referred to as *single tendering* and is, in effect, a special case of selective tendering, the short list containing only one name. It is used when the client has a preference for a particular firm, often because it has done satisfactory work for him before.

Obviously, if only one firm tenders for a job, competition is eliminated and that will, almost inevitably, lead to a higher price. The client may think it is worth paying more, however, in return for a quicker job or one of better quality than he might otherwise get. One should be fairly certain, however, that a worthwhile benefit will accrue before advising a client to nominate rather than go out to competitive tender.

When a contractor is nominated, the contract sum will be arrived at by a process of negotiation. This may be done using bills of quantities or schedules of rates, but instead of the contractor pricing the tender document on his own and submitting his tender to be accepted or rejected, the rates and prices are discussed and agreed until eventually a total price is arrived at which is acceptable to both sides.

Usually the negotiation will be conducted between one of the contractor's senior estimators and the surveyor (either a partner, associate or senior assistant). To facilitate the procedure, one party will usually price the tender document first of all, to provide a basis for the negotiation. The other party will then go through the rates and prices, ticking off those which are acceptable and then the two surveyors will meet to negotiate the unticked ones. When agreement on the whole is reached, a contract will be entered into between the client and the contractor.

Serial tendering

Sometimes, when a large project is to be carried out in successive phases, a combination of selective tendering and nomination is employed. This is sometimes called *serial tendering*. The contractor is chosen for the first phase by means of selective competitive tendering. The accepted tender forms the basis of payment for the resulting contract in the normal way. The tender is also used for the second and later phases, provision being made for so doing in the initial contract by the inclusion of a formula for updating the prices. Alternatively, the contract for the first phase may specify negotiation of new rates, based upon the tendered prices, as the means of determining the payment for each successive phase in the series.

The purpose of serial tendering is to gain the benefits of continuity. The contractor for the first phase of the project will have his site organization set up, his offices, mess and storage huts, etc., already in use and plant of various kinds on the site. When the second phase commences, these facilities will be already available, thus allowing a smooth transition with much less additional expense than if a different contractor were to be employed.

In addition, the contractor's workforce will be familiar with the details of the construction after building the first phase and thus should be able to work more speedily and efficiently on the second and subsequent stages.

Two stage selective tendering

This procedure is used when it is desired to obtain the benefits of competition and at the same time to have the advantage of bringing a contractor into the planning of the project, thus making use of his practical knowledge and expertise. It may also result in an earlier start on site.

The first stage aims to select a suitable contractor by means of limited competition. The second stage is a process of negotiation with the selected contractor on the basis of the first stage tender.

First, a short list of tenderers is prepared, as described for single stage tendering. The *Code* recommends a maximum of six names on the list (four in the case of specialist engineering contracts) and also suggests matters for consideration when drawing up the list.

First stage

When being invited to tender, tenderers are informed of the second stage intentions, including any special requirements of the client and the nature and extent of the contractor's participation during the second stage. Tenderers are asked to tender on the basis of any or all of the following:

(i)   an *ad hoc* schedule of rates, consisting of the main or significant items only;

(ii)   a detailed build-up of prices for the main Preliminaries items;

(iii)   a construction programme showing estimated times and labour and plant resources which would be used, and also construction methods;

(iv)   details of all-in labour rates and main materials prices and discounts which would go into the build-up of the detailed tender;

(v)   percentage additions for profit and overheads;

(vi)   proposed sub-letting of work, with additions for profit and attendance.

During this stage, discussion with each of the tenderers may be conducted in order to elucidate their proposals and to enable the contractors to make any suggestions with regard to design and/or construction methods. When these procedures have been concluded, a contractor is selected to go forward to the second stage. It is important that, in accepting the first stage tender, the parties define procedures for either of them to withdraw should the second stage negotiations prove abortive, and what, if any, payment may be due to either party in that event, including reimbursement of the contractor for any site works the contractor may have carried out.

Second stage

During this stage, finalization of the design proceeds in consultation with the selected contractor, and bills of quantities (or other detailed document describing the proposed works) are prepared and priced on the basis of the first stage tender. Negotiation on the prices will follow until agreement is reached and a total contract sum arrived at, when the parties will enter into a contract for the construction works.

When time is pressing and it is desired to start work on site before final agreement is reached, a contract may be signed earlier. In that case, when the surveyor considers that a sufficient measure of agreement has been reached on the prices for the principal parts of the work, he will recommend the client to proceed with arrangements for the signing of a contract. This is not recommended in the *Code*, but the client may consider that it is worth taking a risk in order to speed up the project.

Tender documents

The number and nature of the tender documents will vary with the type of contract. They will include some or all of the following.

Conditions of the contract

This document sets out the obligations and rights of the parties and the detailed conditions under which a subsequent contract will operate. If a standard form, such as the JCT Form, is used it will not be sent out with the invitation to tender, it being assumed that the tenderers will have a copy or can readily obtain one. The clause headings will, however, be listed in the first (Preliminaries) section of the bills of quantities and/or specification.

Bills of quantities

These are normally used for lump sum contracts of more than say, £50,000. Tenderers should be sent two copies of the bills,[5] one for return to the architect or surveyor with the tender, the other for the contractor to keep as a copy of his submitted prices. If the tenderers are not required to submit priced bills with their tenders, only one copy need be sent to them initially. A second copy will then be sent at the time of requesting submission of the priced bills.

Specification

In the case of lump sum contracts without bills of quantities, a detailed specification will be supplied to tenderers. Sometimes a specification will be supplied in addition to bills of quantities where they are used, but this is exceptional nowadays. A specification may be in a form for detailed pricing.

Schedule of works

As an alternative to a specification in the case of 'without quantities' contracts, tenderers may be supplied with a Schedule of Works. This lists the work comprised in the contract under appropriate headings. The tenderers may be required to price the schedule.

Drawings

Normally general arrangement drawings will be provided, showing site location, position of the building(s) on and means of access to the site and floor plans and elevations. Tenderers are not normally given working drawings as they are not considered to be necessary for pricing purposes, full descriptions of the work being incorporated in the bills or specification. Tenderers are informed, however, that they can inspect drawings not supplied to them, at the architect's office.

Form of tender

This is a pre-printed formal statement, often in the form of a letter, in which a tenderer fills in, in the blank spaces, his name and address and the sum of money for which he offers to carry out the work shown on the drawings and described in the bills of quantities or specification. An example of a typical Form of Tender is given in Appendix C to the *Code of Procedure for Single Stage Selective Tendering*.[6]

Return envelope

Each tenderer should be supplied with a pre-addressed envelope clearly marked 'Tender for (name of project)'. This will ensure that tenders are recognized as such when received and will not be prematurely opened.

Tenderers should be asked to acknowledge in writing receipt of the tender documents.

Tendering procedure

Preliminary enquiry

It is recommended practice (although not always followed) to send, about a month beforehand, to each of the firms from whom it is proposed to invite a tender, a preliminary enquiry[7] to ascertain that they are willing to submit a *bona fide* tender. This avoids the situation of contractors declining to tender or, if they prefer not to decline although not wanting to tender, submitting a 'cover price', i.e. a price which is high enough to be well above the lowest tender.

Sufficient information about the project should be given in the preliminary enquiry letter to enable each contractor to decide whether he is in a position to comply and, if so, whether he would be able in view of other commitments to carry out the contract if it is awarded to him.

If for any reason a firm which has signified its willingness to tender is not included in the final short list of tenderers, it should be informed of the fact immediately, as other tender invitations may be under consideration.

Period for preparation of tenders

It is important that tenderers are given sufficient time to make all necessary enquiries from suppliers, sub-traders, etc. and the date for return of tenders should be fixed so as to allow for the amount of work likely to be involved in preparing a tender for a job of the size and character of the proposed project. Four weeks is recommended as a minimum period, although it is possible, in exceptional cases, for less time to be adequate.

A time of day (often noon) should be specified as the latest time for tenders to be received on the date fixed. Any tenders arriving later should be returned and should be excluded from the competition.

Parity of tendering

It is of paramount importance that all tenders should be based on the same information. Consequently, all the tender documents must be identical. It would be possible, for example, for some copies of a drawing to contain the latest amendments while other copies are of the unamended drawing.

Not infrequently, a tenderer will telephone the architect or surveyor about an item in the bills of quantities or a clause in the specification. He may question the accuracy of a quantity or have been told by a supplier that a specified material is no longer marketed. Such queries must, of course, be answered and must be dealt with promptly. But, whatever the reply given to the enquirer, if it adds to or varies in any way the information given in the tender documents, it must be communicated to all tenderers. This should be done immediately by telephone and then confirmed in writing. The same procedure must be followed if an error is discovered by the surveyor in the bills or specification or other tender documents or if it is decided to extend the time for the receipt of tenders. Tenderers should be asked to confirm in writing the receipt of every written communication of additional or varied information immediately.

Equally in pursuance of the principle of parity of tendering, bills of quantities should be priced as drawn up by the surveyor and likewise where tenders are based on specifications or schedules of rates. A tenderer sometimes wants to price for an alternative material which he considers to be just as good as that specified but which he can obtain more cheaply. Or he may want to allow for an alternative form, method or order of construction. The procedure which he should adopt in such a case is to price the tender documents as printed and to submit, in an accompanying letter or other suitable form, details of the alternative material or form of construction etc. with the consequential effect on his tender sum.

If a tender is qualified in any respect which it is considered has given the tenderer an unfair advantage over the others, or which makes the comparison of the tenders unreasonable, then that tenderer should be given the opportunity of withdrawing the qualification without amending the tender sum.[8] If he is unwilling to do so, his tender should be rejected.

It is desirable that the number of variables in tenders should be reduced as far as possible so that they are made more readily comparable. One of these variables is the length of the construction period and it is common practice for this to be specified by the client.

If a contractor wants to offer to do the work in a shorter time, then the procedure which he should follow is the same as for any other qualification to the printed documents, as described above. If the contractor objects to the period as unreasonably short, however, he should raise the matter with the architect during the tender period. If it is decided to extend the construction period, then all tenderers must be informed, as this may well affect their tender sums.

Opening tenders

Tenders are normally returnable to the architect or project manager (sometimes to the quantity surveyor). A formal procedure should be followed for opening them to eliminate any suspicion of irregularities. No tender must be opened before the latest time for submission and the specially-marked envelopes supplied with the tender documents are intended to eliminate accidental opening. There can be no possibility then of anyone communicating to another tenderer the amount of a competitor's tender.

As little time as reasonably possible should be allowed to elapse before opening tenders and they should all be opened at the same time, preferably in the presence of the architect and quantity surveyor and, if he so wishes, the client. In the case of public bodies, tenders are usually opened by, or in the presence of, the chairman of the committee responsible for the project.

Notifying tenderers

A tender list should be drawn up and sent to all tenderers as soon as possible after the meeting at which the tenders were opened. This is so that each contractor may know whether his tender was successful or not and so be better able to judge his future commitments and know how to respond to any other invitation to tender. The list should contain all the tenderers' names arranged in alphabetical order and all the tender sums in ascending order. It should not disclose which tenderer submitted which amount but, of course, each will be able to identify the position of his own tender in relation to the lowest. At the same time, all but the three lowest should be informed that their tenders have not been successful. The preparation of the tender list and its communication to tenderers is a duty which often falls to the surveyor to carry out.

References

1. *The Placing and Management of Contracts for Building and Civil Engineering Work* (London: HMSO, 1964).
2. NJCC, *Code of Procedure for Single Stage Selective Tendering* (London: RIBA Publications Ltd, 1989. NJCC, *Code of Procedure for Two Stage Selective Tendering* (London: RIBA Publications Ltd, 1983).
3. NJCC, *Code of Procedure for Selective Tendering for Design and Build* (London: RIBA Publications Ltd, 1985).
4. NJCC, *Tendering Procedure for Industrialized Building Projects* (London: RIBA Publications Ltd, 1974).
5. NJCC, *Code of Procedure for Single Stage Selective Tendering* (London: RIBA Publications Ltd, 1989), Appendix B.
6. ibid., Appendix C.
7. ibid., Appendix A.1.
8. ibid., para. 4.4.3.

5

Examining and reporting on tenders

Examination of lowest tender

Once the lowest tender has been identified, the surveyor will proceed to look at it in detail. There will usually be supporting documents such as schedules of rates and, in many cases, bills of quantities requiring careful scrutiny. If there are bills, the lowest tenderer will be asked to submit his priced bills for examination (if not sent with his tender), at the same time being informed that his was the lowest tender and it is therefore under consideration.

To save time in the event of the lowest tender proving unsatisfactory, the next lowest, or the next two lowest tenderers, may also be asked to submit their priced bills. They should be informed that although their tenders were not the lowest, they are under consideration. Opinions vary amongst surveyors as to whether the second or third lowest tenderer's bills should be looked at until the lowest has been rejected. Some say they should not, as being of no concern at all should the lowest tender be accepted. Others take the view that there may be errors in the second or third bills which, if corrected, would make one of them the lowest. From the basic legal point of view that it is the originally tendered total sums which are all-important, it seems logical that, even if the three (or two) lowest tenderers' bills are called for, the second and third ones should not be looked at until the lowest tender has been rejected. Should the lowest be satisfactory, the other bills should be returned to the contractors unexamined.

The surveyor, when examining bills of quantities, schedules of rates, etc., will look for errors of any kind and for any anomalies or pecularities which, in his opinion, make it unwise for the client to enter into a contractual relationship with the tenderer. It should be remembered, however, that whilst it is made clear to tenderers when tendering that the client does not bind himself to accept the lowest or any tender, contractors naturally expect that there will be sound reasons for rejecting the lowest tender. The surveyor must therefore be able to provide those reasons in the event of the lowest tender not being accepted.

It has already been stated that the important factor in a lump sum tender is the total sum for which the contractor offers to carry out the work. If that offer is accepted, then, legally, a binding contract exists. It may be asked therefore 'Why examine the tender in detail?' It is because of the uses made of the component parts of a tender subsequently that any errors in the bills of quantities/ schedules of rates need to be identified and satisfactorily dealt with. Otherwise, problems will almost certainly arise during the course of the contract and in the settlement of the final account. For example, an error may have occurred in a bill when multiplying the quantity of an item by the unit rate and thus an incorrect amount for that item is shown. If the error were not corrected and subsequently the item was affected by a variation order, say, halving the quantity, what is to be done? Should half the total amount be omitted or should the surveyor omit half the quantity at the unit rate shown? The latter course could result in the omission of an amount which is greater than the incorrect total for the item. It is to avoid such difficulties that errors should be found and dealt with in a satisfactory manner.

A further reason is that the examination of the tender details is likely to reveal whether an arithmetical error of such a magnitude has occurred that it would not be wise to hold the contractor to the tender sum because he would then be working at a much reduced profit (assuming the tender sum to be too low). If such an error were in the contractor's favour, then it would be unfair to the client to have to pay more than the job was worth.

The things which the surveyor will look for when examining bills or schedules of rates will be of the following kinds.

Arithmetical errors

These may occur in the item extensions (i.e. the multiplication of the quantities by the unit rates), in the totalling of a page, in the transfer of page totals to collections or summaries, in the calculation of percentage additions on the General Summary or by rounding-off the bill total when carrying it to the Form of Tender.

Pricing errors

These are patent errors *not* matters of opinion as to whether a rate is high or low. A patent error will sometimes occur where, in the transition from cube to superficial measurement, or from superficial measurement to linear, the estimator has continued to use the basic cube rate for the superficial items, or the basic superficial rate for the linear items. Such errors are obvious and not the subject of opinion. Another type of patent error is where, apparently by an oversight, an item has not been priced at all which one would normally expect to be priced.

Again, identical items in different sections of the bills may have been priced differently without there being any good reason. These differences usually arise because the estimator has not remembered that an identical item has occurred before or has not bothered to look back to see what rate had been used. This type of error is not uncommon in housing contracts, where there are separate bills for each house type, and in elemental bills where items recur in different elements.

Pricing method

Sometimes a tenderer will not price most, or any, of the 'Preliminaries' items, their value being included in some or all of the measured rates. This might create difficulty should any adjustment of the value of 'Preliminaries' become necessary.

Some contractors may price all the measured items at net rates (i.e. exclusive of any profit and overheads), showing the latter as an inserted lump sum on the General Summary page of the bills. This is unsatisfactory when using the bill rates later on and the tenderer should be asked to distribute the lump sum throughout the bills or to agree to its being treated as a single percentage addition to all the prices.

Again, a contractor may price the groundwork, in situ concrete and masonry sections of the job (i.e. those sections which will largely be completed in the early months of the job) at inflated prices, balancing these with low rates in the finishings sections. His object would be to secure higher payments in the early valuations, thus improving his cash flow position. This procedure would tend to work against the client's interest, particularly if the contractor were to go into liquidation before the contract was completed. Such a pricing policy might possibly indicate a less stable financial situation than previously thought.

The contractor is entitled, of course, to build up his tender as he thinks fit and cannot be required to change any of his rates or prices. However, contractors usually recognize and accept the need to correct obvious errors and the surveyor would be imprudent not to do so. As regards pricing strategy, this is usually considered to be a matter for the contractor's decision, the remedy being, if the surveyor has genuine grounds for concern about it, for him to recommend the client not to accept the tender.

Basic prices list

The contract provisions may require the contractor to submit a list of the basic prices of materials upon which his tender is based. This will be used in order to adjust the contract sum for any fluctuations in the costs of materials during the contract period (see pp. 143 and 341). The list should be scrutinized to ensure that the prices quoted are the current market prices of the materials named, including delivery to the site in full loads. If the surveyor considers it necessary, he should ask the contractor to submit recent quotations for each of the listed materials.

Further points to note in regard to tenders which are not based on bills of firm quantities are as follows.

Tenders based on bills of approximate quantities

The rates for items of which the quantities are likely to be substantially increased upon measurement of the work as executed should be carefully scrutinized, as these may affect the total cost of the job considerably.

Tenders based on drawings and specification only

The schedule of the principal rates which have been used in the preparation of the tender should be looked at closely as these rates will be used in the pricing of any variations which may be subsequently ordered. If there is likely to be much dayworks, the percentage additions to prime costs which the tenderer has quoted may also become significant.

Tenders based on ad hoc schedule of rates

The rates for items which are likely to be significant in terms of quantity need careful scrutiny. The above comment in regard to dayworks will also apply here.

Reporting on tenders

The surveyor must report to the project manager or the architect and the client as soon as his examination of the tenders is complete. He must remember that the purpose of the report is to enable the client to decide whether to accept any of the tenders and, if so, which one. Consequently, it should concentrate on matters of importance and minor matters, such as the details of arithmetical errors, should be excluded.

The form of the report will vary according to the nature of the tender documents but usually will include the following:

(*a*) the opinion of the surveyor as to the price level, i.e. that the tender is high, low or about the level expected;
(*b*) the quality of the pricing, indicating any detectable pricing method or policy;
(*c*) the extent of errors and inconsistencies in pricing and the action taken in regard to them;
(*d*) the details of any qualifications to the tender;
(*e*) the likely total cost of the project, if not a lump sum contract;
(*f*) a recommendation as to acceptance or otherwise.

The examination and reporting on tenders for *design and build contracts* (see p. 30) is not as straightforward a matter as for other types of contract. There will be a number of variables, making comparison between tenders difficult. Apart from price differences, they will vary in the construction time, constructional form, quality of finishes, degree of maintenance likely to be required subsequently, and perhaps the anticipated cost of running and maintaining specialist installations. In his report, the surveyor will need to consider all the tenders since there will not be a lowest tender as such. He will need to set out in a clear and easily-assimilated form the relative savings, benefits, merits and demerits between the packages offered. He should, nevertheless, make his recommendation as to which one, if any, the client should accept.

Figure 5.1 is a report on the tenders for the contract described in Appendix A and illustrates the application of some of the foregoing principles.

Dealing with errors, etc.

In the case of tenders not based on bills of quantities, the scope for detection of errors, etc. is limited and correction of any which are discovered is usually a straightforward matter of adjusting any erroneous rates, with the contractor's agreement.

If a lump sum tender based on drawings and specification only shows a breakdown of the total and this contains arithmetical errors, then correction should be done in a similar way to that described for bills of quantities, so far as is applicable.

Two alternative ways of dealing with errors in bills are recommended in Section 6 of the *Code of Procedure for Single Stage Selective Tendering*.[1] The first is the traditional procedure of informing the tenderer of the errors and asking him whether he wishes to confirm his tender, notwithstanding the errors (which may be favourable or unfavourable to him), or whether he wishes to withdraw it. This procedure accords with the basic principle of the law of contract whereby an offer is made to provide goods and/or services for a certain consideration and the offerer may withdraw his offer at any time before its acceptance.

The second course of action is to inform the tenderer of the errors and to ask him whether he wishes to confirm his tender, notwithstanding the errors, or whether he wishes to correct genuine errors. If he chooses to correct his errors and the revised

KEWESS & PARTNERS
52 High Street, Urbiston, Middlesex, UN2 1QS.

SHOPS AND FLATS, THAMES STREET, SKINTON, MIDDLESEX

Report on Tenders

1. Tender list. Six tenders were received as follows:

Contractor	Tender Sum
Beecon Ltd	£495,000
W.E. Bildit Ltd	£514,800
E.A. Myson Ltd	£520,000
Thorn (Contractors) Ltd	£526,500
Smith & Sons Ltd	£538,200
Brown & Green Ltd	£576,000

The lowest tender is 3.85% below the second lowest tender. It is also within the amount of our last amended cost plan dated 20 November, 1993, which was in the sum of £504,000.

2. Bills of Quantities. The priced bills of quantities submitted by the lowest tenderer, Beecon Ltd., have been examined and the following is a report on our findings.

3. Pricing.
3.1 The general level of pricing is, in our opinion, slightly lower than average for this class of work, even allowing for the present economic climate. We do not consider, however, that this should be a cause for concern in this instance as Beecon Ltd. are well-established contractors in the area with a reputation for good quality work. Also, it is known that they particularly wanted to obtain this contract because of its prominent location and its proximity to their head office.

3.2 The pricing is consistent throughout with no serious anomalies or detectable adverse characteristics.

3.3 A number of arithmetical errors were discovered and have been corrected in accordance with the provisions set out in the tender document. Beecon Ltd. have elected to stand by their tender in the sum of £495,000 notwithstanding the errors. Had the tender sum been amended, it would have been reduced by £9,112, which is the net amount of the errors.

4. Recommendation. We recommend that the contract be awarded to Beecon Ltd. in consideration of the sum of £495,000.

6 February, 1994.

Figure 5.1 Report on tenders

tender sum is then higher than the next on the list, then his tender will be set aside and the one now lowest will be proceeded with in a similar way instead. A recommendation similar to this second alternative is contained in Section 7 of the *Code of Procedure for Two Stage Selective Tendering*[2] in relation to first stage tenders.

It should be noted that the term 'genuine errors' means inadvertent arithmetical errors and patent pricing errors, as defined earlier. It does not include errors which the estimator may have made in the build-up of a bill rate or price. The provision for the correction of genuine errors does not give the contractor the opportunity of increasing any rates or prices which on reflection he realizes are too low nor of pricing any items which were intentionally left unpriced.

One reason why the second alternative has been introduced is that, because of the large sums involved in most construction projects, it is in clients' interests to allow the correction of such errors as have the effect of increasing the tender sum, if to do so leaves the tender still in the lowest position. The contractor will be happier and the client will be paying a lower price than if the contractor felt he had no alternative but to withdraw. In any case, it is not in the client's interest to have a discontented contractor from the start of the contract, who feels that he is beginning the contract with an expectation of a reduced profit.

The *Code* recommends that it is decided from the outset which of the alternative methods of dealing with any errors will be used and that this information is included in the preliminary enquiry to tenderers, the formal invitation to tender and the Form of Tender.

When the contractor elects to confirm his tender, his errors must still be dealt with so that the priced document is arithmetically correct and consistent throughout. This is so that when the rates and prices are used subsequently, difficulties which would otherwise arise will be eliminated (see p. 58 under the heading '*Examination of lowest tender*'). The correction of errors must, of course, be done in consultation and agreement with the contractor. The surveyor will normally draw up a list of errors, etc., which he will agree with the contractor and the surveyor will then amend the priced document accordingly.

The procedure for the correction of errors in bills is as follows:

(*a*) If the original text has been altered without authorization, the alteration(s) should be deleted and prices adjusted accordingly.

(*b*) All errors of the kinds described earlier (i.e. arithmetical and patent pricing errors) should be corrected and the corrections carried through from page totals to collections, from collections to summaries and from summaries to general summary. The incorrect amounts should be ruled through with a single line and the correct ones written above, neatly and clearly.

(*c*) A sum, equal to the net amount of the aggregated errors, should be inserted on the General Summary immediately before the final total as an addition or deduction, as appropriate. This will have the effect of adjusting the corrected bill total to the amount stated on the Form of Tender (which, it should be noted, in some instances may not be the same sum as the original total of the bills).

This net addition or deduction equal to the errors is, in effect, the aggregate of adjustments to every rate and price in the bills, excluding any contingency sum, provisional and prime cost sums and Preliminaries (if deducted). Accordingly, any subsequent use of a rate or price from the bills or of one based upon rates or prices in the bills, must be subjected to an equivalent adjustment. To facilitate this, the balancing sum (i.e. the net amount of the errors) is calculated as a percentage of the value of the builder's work. The contract copies of the priced bills should be endorsed with a statement that all the rates and prices (other than contingency, provisional and prime cost sums) are to be considered as increased or reduced (as appropriate) by that percentage. The example in Figure 5.2 illustrates how the process may be effected.

All surveyors do not agree on what constitutes 'builder's work' for the purpose of calculating the adjusting percentage referred to in the last paragraph. The *Codes* say that in addition to the fixed sums mentioned above, the value of Preliminaries items should also be excluded. Turner[3] does not agree, taking the view that the latter are just as much the contractor's own prices as those applied to measured items. This seems a reasonable view, particularly as arithmetical errors may occur just as easily in the Preliminaries as in the measured sections of the bills.

LIST OF ERRORS IN PRICED BILLS SUBMITTED BY BEECON LTD.

Page no.	Item ref.	Type of error	Effect of error on total of tender	
			Increased by £	Decreased by £
25	–	Page total	100.00	
30	–	Page total		6.00
48	C	Item extension	65.20	
79	–	Page total transferred incorrectly		0.45
86	L	Item extension		18.50
103	D	Item extension	460.00	
140	–	Page total		500.00
General Summary	–	Page total transferred incorrectly from p.137		27.00
General Summary		Sub-total	10 000.00	
Form of Tender		Total rounded down		961.25
			10 625.20	1 513.20
			1 513.20	
		Net amount of errors	9 112.00	

The following pages show how the error in item D on p.103 of the bills was dealt with and the consequential correction of totals, also the corrections to the Summary and General Summary of all the errors listed above. On the General Summary is shown the sum added to balance the net error and also the endorsement. The adjusting percentage referred to in the endorsement has been calculated thus:

$$\frac{9112.00}{485\ 888.00 - (35\ 660.00 + 117\ 724.00 = 5820.36)} \times 100 = 2.79\%$$

This calculation should be shown on the General Summary page in a convenient position or on the facing page.

Figure 5.2 Example of correction of errors in bills of quantities

```
L40 GENERAL GLAZING

STANDARD PLAIN GLASS                                              £    p

OQ quality sheet glass

A 4mm Panes, area 0.15-4.00m² to wood with )
  bradded wood beads.                        ) 71   m²  27·85   1977 35

B 4mm Ditto but bedded in glazing compound.) 60   m²  25·98   1558 80

C 4mm Ditto, area not exceeding 0.15m²       )
  (In 110 nr.panes).                         ) 16   m²  29·01    464 16

D 4mm Ditto, area 0.15-4.00m² to metal       )              1195 08
  with mastic and spring glazing clips.      ) 46   m²  25·98   1655 08

  Rolled glass, pattern group 2

E 3mm Panes, area 0.15-4.00m², to wood       )
  with bradded wood beads.                   ) 20   m²  25·13    502 60

  Rolled Georgian wired rough cast glass

F 6mm Panes, area 0.15-4.00m², to wood       )
  with bradded wood beads.                   ) 48   m²  42·25   2028 00

  Rolled Georgian wired polished plate
  glass

G 6mm Panes, area 0.15-4.00m², to wood       )
  with bradded wood beads                    ) 17   m²  54·53    927 01

H 6mm Ditto, area not exceeding 0.15m²,      )
  ditto (In 140 panes).                      ) 20   m²  54·86   1097 20

  GG quality clear plate float glass

J 6mm Panes in sliding hatch doors, 475mm    )
  wide x 570mm high, ground on all edges,    )
  1 nr. ground and shaped finger grip        )
  sinking.                                   ) 20   nr  28·57    571 40
                                                              ─────────
                                                              10321 60
                     Carried to Collection                  £10781 60

                  103
```

Figure 5.2 continued

L40 GENERAL GLAZING (CONT)

				£	p
STRIPS FOR EDGES OF PANES					
A Glazing strips as described to edges of 3mm glass.	) 188	m	1.08	203	04
B Ditto but 4mm glass.	) 320	m	1.08	345	60
C Ditto but 6mm glass.	) 406	m	1 08	438	48

Carried to Collection   £   987 12

COLLECTION		£	p
Page	93	2 560	20
"	94	825	74
"	95	1 317	09
"	96	2 150	65
"	97	5 792	20
"	98	1 746	51
"	99	3 107	26
"	100	2 329	15
"	101	2834	21
"	102	2 244	77
"	103	10 321 60 ~~10 781 60~~	
"	104	987	12

WINDOWS/DOORS/STAIRS Total Carried to Summary   £   36 216 50 ~~36 676 50~~

Figure 5.2 continued

```
                    BILL  NO  3  -  SHOPS  AND  FLATS

                              SUMMARY


PAGE                                                          £      p
                                                          3i  256   57
   26      GROUNDWORKS                                    -31- 356- -57-

                                                          54  488   22
   33      IN  SITU  CONCRETE                             -54- 482- -22-

   44      MASONRY                                        45  417   33

                                                          20  342   64
   69      STRUCTURAL/CARCASSING  METAL/TIMBER            -20- 407- -84-

                                                          9   245   12
   79      CLADDING/COVERING                              -9- 244- -67-

                                                          18  587   66
   92      WATERPROOFING                                  -18- 569- -16-

                                                          36  216   50
  104      WINDOWS/DOORS/STAIRS                           -36- 676- -50-

  109      SURFACE  FINISHES                              15  100   50

  112      FURNITURE/EQUIPMENT                            3   819   96

  116      BUILDING  FABRIC  SUNDRIES                     9   256   15

  124      DISPOSAL  SYSTEMS                              8   326   16

  129      PIPED  SUPPLY  SYSTEMS                         19  149   60

  137      ELECTRICAL  SERVICES                           9   389   72


                                                         280  596   13
            TOTAL  CARRIED  TO  GENERAL  SUMMARY    £ -281- 196- -38-



                              137
```

Figure 5.2 continued

```
                        GENERAL  SUMMARY

PAGE                                                           £     p

  15    BILL  NO  1 -  PRELIMINARIES/GENERAL  CONDITIONS     35 660   00

  19    BILL  NO  2 -  DEMOLITION                             5 531   00
                                                           280 596   13
 137    BILL  NO  3 -  SHOPS  AND  FLATS                   -281-169- -38-
                                                            40 556   51
 143    BILL  NO  4 -  EXTERNAL  WORKS                      -40-056- -51-

 147    BILL  NO  5 -  PRIME  COST  AND  PROVISIONAL  SUMS  117 724   00


                                                           480 067   64
                                                          -490-140- -89-

        Add for:

        Insurance against injury to persons
        and property as item 5C                              3 632   00

        All risks insurance as item 5C                         811   36

        Water for the Works as item 13A                      1 377   00
                                                           485 888   00
        Add to adjust for errors                             9 112   00
                                                           495 000   00
        TOTAL CARRIED TO FORM OF TENDER                    -495-961- -25-*
```

All rates and prices herein (excluding preliminaries,
prime cost and provisional sums) are to be
considered as increased by 2·79 %

Signed. . . . .   Signed. . . . .

148

Figure 5.2 continued

* Carried to Form of Tender as £495 000

In the calculation on page 66 the Preliminaries have been deducted from the contract sum in accordance with the *Codes*. The bills were those submitted in support of the tender for the contract described in Appendix A, in which the corrected amounts are shown in the Summary of Bill No 3 and in the General Summary.

References

1. NJCC, *Code of Procedure for Single Stage Selective Tendering* (London: RIBA Publications Ltd, 1989).
2. NJCC, *Code of Procedure for Two Stage Selective Tendering* (London: RIBA Publications Ltd, 1983).
3. TURNER, DENNIS F., *Quantity Surveying Practice and Administration* (London: George Godwin Ltd, 3rd edn., 1983), p. 125.

6

The contract

Under the general law of contract, when a party makes an offer to provide goods and/or services for some certain consideration and the party to whom the offer is made accepts it, then, provided it does not involve any illegal act, a contract which is enforceable at law exists. This is no less the case in the construction world than in any other sphere of business or industry. The offer is made by a contractor who tenders to carry out specified construction works in return for a money payment and upon the acceptance of that offer by the client promoting the project, a binding contract comes into being.

It is desirable that the client's acceptance should be in writing and that it should be given as soon as possible after receipt of tenders. If there is any appreciable delay of, say, one to two months or more, then it will be necessary to obtain the contractor's confirmation that his price still stands or what the increase in his price is, due to rising costs.

The contract documents

The number and nature of the contract documents will normally correspond with those of the tender documents (see p. 51). They may be some or all of the following.

Form of contract

This is the principal document and will often be a printed standard form, such as one of the variants of the JCT Form. This consists of three parts: (i) the Articles of Agreement, which is the actual contract which the parties sign; (ii) The Conditions (sub-divided into five parts), which set out the obligations and rights of the parties and detail the conditions under which the contract is to be carried out; and (iii) Supplemental Provisions (the VAT Agreement), which sets out the rights and liabilities of the parties in regard to value added tax.

Bills of quantities

Any errors in the bills must be corrected in the manner described in Chapter 5 and any necessary adjustments to rates and prices must be clearly and neatly made.

Specification

Under the 'with quantities' variants of the JCT Form, the specification (if one exists) is not a contract document but it is under the 'without quantities' variants. In the latter case, the specification may have been prepared in a form for detailed pricing and have been priced by the contractor when tendering. If this is not the case, the contractor may have been asked to submit with the tender a Contract Sum Analysis. This shows a break-down of the tender sum in sufficient detail to enable variations and provisional sum work to be valued, the price adjustment formulae to be applied and the preparation of interim certificates to be facilitated.

JCT Practice Note 23[1] gives a guide to the identification of specified parts of the contract sum and the breakdown of the remainder into parcels of work. A Contract Sum Analysis may not be required where a Schedule of Rates is used. In cases where a Contract Sum Analysis or a Schedule of Rates is provided, it is signed by the parties and attached to the contract, but is not a contract document. However, the specification is a contract document in such cases.

Schedule of Works

Where bills of quantities do not form part of the contract (for example, under the JCT Standard Form without quantities), the contractor may be sent a Schedule of Works for pricing when tendering, instead of a specification. If so, the priced Schedule of Works will become a contract document.

Schedule of rates

Where bills of quantities are not provided, a schedule of rates is usually necessary as a basis for pricing the work in measured contracts and for pricing variations in the case of lump sum contracts. Such schedules are described on pp. 15–16.

Drawings

The contract drawings are not limited to those (if any) sent to the contractor with the invitation to tender but are all those which have been used in the preparation of the bills of quantities or specification.

It is important that the contract drawings are precisely defined, as amendments may be made to them during the period between inviting tenders and commencing the work on site. Such amendments may affect the value of the contract and, if they do, should be made the subject of Architect's Instructions. It is good practice therefore, for the surveyor to certify in writing, on each of the drawings used for taking-off or specification purposes, that they were the ones so used.

Preparations for executing the contract

The surveyor is often regarded by architects as the expert in matters relating to the interpretation and application of the contract Conditions. Accordingly, it is not uncommon for the preparations for the signing of the contract to be the task of the surveyor.

Each of the contract documents is important from a legal point of view. Both parties (and their agents) are bound by what is said in them. When a contract proceeds satisfactorily to completion and settlement, many of the details in those documents may not be referred to. They become all-important, however, when a contract runs into difficulties and particular statements in the documents may then come under close scrutiny. If there are ambiguities, discrepancies or contradictions in them, it may lead to delays, arbitration proceedings or even actions in the Courts, all of which are expensive as well as frustrating.

If such faults in the contract documents are due to the carelessness of the persons responsible for their preparation, their principals (i.e. the partners in the firm of surveyors or architects, as the case may be) may be liable for a claim for damages for negligence, which might prove very costly. Great care must be exercised therefore, in carrying out the task of preparing the documents for the execution of the contract.

Each of the contract documents as already listed will require attention, if only to facilitate the actual signing, and they will now be considered in turn.

Form of contract

The Articles of Agreement

The blank spaces must be filled in with the date and names and addresses of the Employer and the contractor, a brief statement of the nature and location of the Works, the architect's name, the numbers of the contract drawings, the contract sum (in words and in figures) and the names and addresses respectively of the architect/contract administrator and quantity surveyor.

Where bills of quantities are not used and if, in consequence, a quantity surveyor is not appointed, an alternative Article is provided for the name to be given of the person exercising the quantity surveying functions. The Article not required should be struck out in a neat and clear manner.

In the 'local authorities' variants of the JCT Form, an alternative Article (3B) is given for use where the architectural functions are to be exercised by a contract administrator, being a person who is not a registered architect and again, the Article not required must be struck out.

There follow in the 'private' editions of the JCT Form, spaces for the signatures of the parties. The Employer (or his representative) signs in the first space and the contractor (or his representative) signs in the second space. Each signature is followed by the signature of a witness. If both signatures are obtained at the same time, as is usual, the witness may be the same person in each instance.

When either party must, or wishes to, execute the contract as a deed, as in the case of some local authorities and other public bodies and corporations, the spaces for signatures, referred to above, are left blank and the seals are affixed in the appropriate spaces indicated.

The 'local authorities' editions of the JCT Form have a blank page for the insertion of a suitable Attestation Clause in place of the printed clauses in the 'private' editions.

The Conditions

Some clauses of the Conditions in the JCT Form contain alternative wording. Attention is drawn to these alternatives in footnotes and to the need for appropriate deletions. Such deletions should be clearly and neatly made.

There may be need or desire for other alterations or additions to be made to the clauses of a standard form in a particular case. All such alterations and additions, as well as the alternatives to be deleted, should be specifically listed in the 'Preliminaries' section of the bills or specification and, if this has been done, the list may then be used as a check-list in the preparation of the standard form for executing the contract.

The blank spaces in the Appendix to the Conditions (see Appendix B, p. 347) should also be filled in, in accordance with a further list in the bills or specification. It should be noted that if nothing is stated against certain of the clause numbers, then specified time periods will apply.

The parties should initial each amendment to the printed document in the adjacent margin. This serves to confirm that all amendments were made before the contract was executed and that no amendment was made subsequent to that event.

Bills of quantities

The priced bills submitted by the contractor should show clearly all the amended and corrected rates and prices, etc., which have been agreed between the contractor and the surveyor. If an adjusting amount, equal to the amount of the errors, has been entered in the General Summary, then an endorsement should be added indicating the percentage effect on the measured rates and prices (see p. 70). On the reverse of the last page a further endorsement should be added, saying 'These are the Contract Bills referred to in the Contract' with spaces for the signatures of the parties, set out in the same way as in the Articles of Agreement.

Specification

Where the specification is a contract document, all that is necessary is to endorse on the reverse of the last page the words 'This is the Specification referred to in the Contract' with spaces for the signatures as in the case of bills. The schedule of rates should show clearly all corrections and amendments which have been agreed between the surveyor and the contractor. If, alternatively, a Contract Sum Analysis has been submitted, this should be signed by the parties and attached to the Specification or to the Contract Conditions.

Schedule of Works

If a priced Schedule of Works has been submitted by the contractor when tendering, that will be a contract document instead of the specification. It should be endorsed in a similar manner as that described for specifications and be signed by the parties.

Drawings

It is desirable that each of the drawings should be endorsed either 'This is a contract drawing' or 'This drawing was used in the preparation of the bills of quantities' or some similar statement. Many firms have rubber stamps for this purpose.

Subsequent to the execution of the contract, the contract documents are kept either by the architect or by the surveyor. The Employer and the contractor have the right to inspect them at any reasonable time. (The 'local authorities' variant provides for these documents to remain in the custody of the Employer.)

Copies of contract documents

Immediately after the execution of the contract, the contractor should be given a certified copy of the Articles of Agreement, the Conditions of the Contract, the contract drawings, and the contract bills. In practice, the two copies of these documents are prepared at the same time so that they are identical, including all the initialling and the signatures. Where the contract is sealed, only one copy needs to bear the seals.

The contractor should be given in addition, two copies of the unpriced bills, or, if there are no bills, two copies of the specification, and two copies of the drawings, plus two copies of any descriptive schedules or similar documents. If the surveyor is responsible for organizing the arrangements for the signing, he will usually collect together beforehand two sets of the documents and hand them to the contractor at the meeting at which the contract is executed.

The contract sum

In the case of a lump sum contract, a specified sum of money is stated in the contract as to be paid by the Employer to the contractor for the carrying out of the construction work. This sum is seldom, if ever, a fixed amount but is subject to additions and deductions in respect of various matters. Consequently, the JCT Form, in the Articles of Agreement, after providing for the insertion of the contract sum, continues 'or such other sum as shall become payable hereunder at the times and in the manner specified in the Conditions'.

The main causes for adjustment of the contract sum for which the Conditions make provision are: the adjustment of prime cost and provisional sums; variations to the design and/or the specification of the work; additions or reductions to the scope of the work, loss or expense incurred by the contractor for specified reasons and increases or decreases in the costs of labour and materials or in taxes, levies or contributions imposed by Government.

The variable nature of the contract sum may, at first sight, appear to nullify the value of stating any sum at all. It does have the merit, however, of giving the parties a good indication of the level of cost and providing a basis for estimating the eventual cost at later stages in the progress of the project.

The scope for adjustment of the contract sum is not unlimited. Clause 14.2 of the JCT Form limits adjustment to the express provisions of the Conditions of the contract. It specifically excludes the correction of errors made in the contractor's computation of the contract sum.

Reference

1. JCT, *Practice Note 23, A Contract Sum Analysis* (London: RIBA Publication Ltd, 1987).

Bibliography

1. KEATING, D., *Keating on Building Contracts* (London: Sweet & Maxwell, 6th edn, 1995).

7

Prime cost and provisional sums

Definitions

Prime cost (p.c.) sums and provisional sums are lump sums included in the contract sum to cover the cost of parts of the work which are not measured in the bills (in the case of a 'with quantities' contract) or specified in detail in the specification (in the case of a 'without quantities' contract).

A *prime cost sum* is a sum provided for work or services to be executed by a nominated sub-contractor or for materials or goods to be obtained from a nominated supplier.

A *provisional sum* is a sum provided for work or services to be executed by a statutory authority or a statutory undertaking, or for either 'defined' or 'undefined' work. *Defined work* is work which is not completely designed at the time the tender documents are issued but for which certain specified information can be given.[1] *Undefined work* is work for which such specific information cannot be given.

The right to nominate is solely the architect's, who may exercise that right

(a) by naming a *sub-contractor* or *supplier* in a prime cost sum item in the bills of quantities or specification;

(b) by naming a *supplier* in an architect's instruction relating to the expenditure of a prime cost sum or a provisional sum;

(c) by naming a *sub-contractor* on the Standard Form of Nomination Instruction for a Sub-Contractor NSC/N to carry out work for which a prime cost sum is included in the bills of quantities or specification.

The reasons for dealing with parts of the work by the use of p.c. sums are:

(a) in the case of nominated sub-contractor's work, to give the architect control over the choice of firm to carry out the work, particularly where it is of a specialist nature or outside the scope of work normally carried out by general contractors;

(b) in the case of nominated suppliers' materials or goods, to enable the architect to select articles or materials manufactured by particular firms.

A special kind of provisional sum is one for contingencies, known as a *contingency sum*. This is to meet or offset costs of undefined work which cannot be foreseen before construction begins and which may not arise at all. A contingency sum has no real relation therefore to the contract works, and in practice merely serves to reduce the total net cost of any extra or more expensive work than that originally envisaged.

Sub-contracts and contracts of sale

In order to safeguard the rights of the parties in a building project and to make their respective obligations legally binding, a contractor is required to enter into sub-contracts and contracts of sale with nominated sub-contractors and nominated suppliers respectively. So, in relation to sub-contract work, clause 35.4 lists standard forms of tender, sub-contract and employer/sub-contractor agreement and nomination, the use of which constitutes a standard procedure described in clauses 35.5 to 35.9.

The provisions in clause 36 (Nominated Suppliers) are much simpler. They specify various rights and obligations to be included in a contract of sale, unless the architect and contractor agree otherwise.

It is important, therefore, that the quantity surveyor when dealing with prime cost sums is aware of the agreements which have been entered into by the contractor and must adjust the contract sum accordingly, within the scope and limitations of those agreements.

Prime cost sums

The term *prime cost* has been defined already in Chapter 2 to mean 'the total cost to the contractor of buying materials, goods and components, of using or hiring plant and of employing labour'. In the present context, its meaning is extended to include the cost to the contractor of employing nominated sub-contractors to carry out specified parts of the Works, and the cost of materials and goods obtained from nominated suppliers.

Clauses 35 and 36 of the JCT Form describe the manner in which prime cost sums should be dealt with. Put simply, when the contract sum is adjusted at the end of the contract, the prime cost sums are deducted and the corresponding actual sums expended are substituted. In practice, however, various matters of detail are involved, which must be dealt with properly. These are:

(*a*) discounts;
(*b*) net prices and incorrect discount;
(*c*) general contractor's profit;
(*d*) attendances on nominated sub-contractors.

Discounts

Discounts are amounts by which quoted or stated costs or prices may be reduced when paying accounts, usually depending on the fulfilment of specified conditions.

(*i*) Trade discounts

These discounts are customarily allowed off the prices on standard price lists of materials and goods in particular trades. This is often done to avoid frequent revision of such lists, the discount being varied to allow for fluctuations in prices. In addition, some suppliers will allow larger discounts to regular customers or those who buy large quantities. Contractors may thus obtain more favourable discounts from one merchant than another.

(*ii*) Cash discounts

These are allowed off invoice prices if the contractor pays within a certain period. In the case of sub-contractors, this period is seventeen days after issue of the architect's certificate which includes the value of the invoiced work or services. In the case of suppliers, the period is thirty days from the end of the month during which delivery of the materials or goods is made. If payment is not made within the specified period, the contractor is not entitled to deduct the discount. As far as the surveyor is concerned, however, the amount allowed against the p.c. sum will always include the appropriate cash discount (but see below under 'Net prices and incorrect discount').

Under the JCT Form, all discounts *except* cash discounts, must be deducted from any sums due to sub-contractors and suppliers. In addition, cash discounts in excess of $2\frac{1}{2}$ per cent in the case of nominated sub-contractors and 5 per cent in the case of nominated suppliers must also be deducted. This means that the Employer has the benefit of all discounts other than the specified percentages for cash discounts.

Net prices and incorrect discount

Sometimes sub-contractors and suppliers quote net prices (that is, prices exclusive of cash discount) or prices inclusive of cash discount but at the wrong percentage. The provisions of the Standard Forms give the contractor the right to the appropriate discount and the Employer the right not to have to pay more than the appropriate discount. They also, in effect, prevent the architect nominating a sub-contractor or supplier who will not allow such discount.

The use of the JCT Standard Form of Nominated Sub-Contract Tender NSC/T ensures that the discount included in sub-contractors' tenders is correct as it refers in Part 2, page 3, to the cash discount allowable to the Main Contractor under Sub-Contract NSC/C clause 4.16.1.1, i.e. $2\frac{1}{2}$ per cent. However, in cases where the standard procedure is not followed, difficulties in regard to discount may arise and consideration needs to be given to the proper way to deal with them. The following paragraphs relate to the correction of amounts where the proper discount has not been included in tenders or in invoices for sub-contractors' work or suppliers' goods.

The simple solution where quotations or tenders are net or include the wrong percentage is for the architect to request the sub-contractor or supplier to revise the quotation to include the correct discount. They will usually do so without demur.

If the architect instructs the contractor to place an order against a quotation which includes incorrect discount or no discount at all, the contractor should draw the architect's attention to the error before proceeding. If the contractor places the order without query, he may be deemed to have waived his right to the discount altogether.[2]

It has been argued[3] that, under the JCT Form, if the architect so instructs the contractor then he is acting outside his powers and that the proper course for the contractor is to request the architect, under clause 4.2, to specify under what clause he has issued the Instruction. This course would probably lead to clause 13 being used to provide the due compensation for loss of discount.

In practice, however, things are not always done 'by the book'. The architect may not think it to be worthwhile asking for a revised quotation and the contractor, on his part, may not wish to appear to be challenging the architect. The latter may therefore say to the contractor, 'The surveyor will allow for the discount in the final account' and will request the surveyor so to do. The contractor will usually be disposed to accept this solution to the difficulty. The surveyor must then make the appropriate adjustment to the invoice amount when adjusting the p.c. sum.

It should be noted that the effect of dealing with the matter in this way is that the contractor receives a sum equivalent to the discount irrespective of the time of payment of the invoice.

The following examples show the calculations in the more common cases. Each assumes that the architect instructed the contractor to place an order on the basis of a quotation which included either the wrong discount or no discount, promising that the surveyor would adjust for the error in the final account.

Example 1

A nominated sub-contractor rendered an invoice for £6240 for floor tiling supplied and laid, the invoice being marked 'net, monthly account'.*

---

* The words 'monthly account' indicate that the credit agreement between contractor and sub-contractor is that credit is given on a monthly basis only.

The invoice ought to have included 2½ per cent cash discount and to correct the error, ⅟₃₉th should be added, because the percentages are always deductible from gross (i.e. inclusive) totals.

		£
Correction: Amount of invoice		6240.00
Add ⅟₃₉th		160.00
Corrected total		6400.00

Check: 2½% of £6400.00 = £160.00.)

The sum of £6400.00 will be paid by the Employer to the contractor who will retain 2½ per cent of that sum (i.e. £160.00) and pay the balance of £6240.00 to the sub-contractor.

Example 2

A nominated sub-contractor rendered an invoice for £12,480.00 for an electrical installation. The invoice stated '5 per cent cash discount, monthly account'.

The percentage was incorrect and should have been 2½ per cent. The Employer is entitled to receive the benefit of discounts greater than 2½ per cent; consequently, the difference must be deducted to arrive at the proper sum to be paid to the contractor, who may then deduct 2½ per cent as the contractor's discount. To correct the total, the incorrect 5 per cent should be deducted and an addition of ⅟₃₉th made to the resulting net amount.

		£
Correction: Amount of invoice		12 480.00
Deduct 5%		624.00
		11 856.00
Add ⅟₃₉th		304.00
Corrected total		12 160.00

(*Check:* 2½% of £12 160.00 = £304.00.)

The Employer will pay the contractor £12,160.00 who will then deduct 2½ per cent for cash discount and pay the sub-contractor £11,856.00, which equals the invoice amount less the 5 per cent shown on the invoice.

Example 3

A nominated supplier of ironmongery invoiced the goods supplied at £6080.76 and marked the invoice 'net'.

Cash discount should have been included at 5 per cent. $\frac{1}{19}$th should be added in this case to correct the error.

		£
Correction:	Amount of invoice	6080.76
	Add $\frac{1}{19}$th	320.04
	Corrected total	6400.80

(*Check*: 5% of £6400.80 = £320.04.)

Example 4

A nominated supplier of sanitary appliances invoiced goods at £9135.20, the invoice showing cash discount at $2\frac{1}{2}$ per cent instead of 5 per cent. To correct the error, the incorrect $2\frac{1}{2}$ per cent should be deducted and 5 per cent substituted by adding $\frac{1}{19}$th to the net amount.

		£
Correction:	Amount of invoice	9135.20
	Deduct $2\frac{1}{2}$%	228.38
		8906.82
	Add $\frac{1}{19}$th	468.78
	Corrected total	9375.60

(*Check*: 5% of £9375.60 = £468.78.)

The Employer will pay the contractor £9375.60 who will then deduct the rightful discount of 5 per cent (£468.78) and pay the supplier the balance of £8906.82, which is the sum the latter will expect to be paid having allowed $2\frac{1}{2}$ per cent off the sum invoiced.

Clause 4.16 of the JCT Nominated Sub-Contract Conditions (NSC/C) gives the contractor the right to deduct $2\frac{1}{2}$ per cent cash discount from *any* sum due to a sub-contractor which is included in the amount of a certificate for payment, if payment is made within the stipulated period of seventeen days. This has sometimes been interpreted as giving the contractor the right to deduct the discount from a net sum shown on an invoice rendered by a sub-contractor, with whom an order had previously been placed on the basis of a net quotation.

It should be remembered, however, that those provisions of clause 4.16 which relate to the deduction of cash discount are based on the premise that the architect will have nominated the sub-contractor on the basis of a tender submitted on the JCT Form of Nominated Sub-Contract Tender (NSC/T), which specifically refers to allowance made for the discount. Thus the clause does not apply to the situation where a net quotation has been accepted. In that circumstance, an adjustment of the kind illustrated above would need to be made in order that the provisions of the clause may become effective and the contractor receive the benefit of the discount to which he is entitled.

It should be noted that the stipulated percentages for cash discount apply to the total cost of materials or goods supplied by a nominated supplier (including any tax or duty and the cost of packing, carriage and delivery), and to the total value of sub-contract works (including any dayworks and fluctuations amounts).

Clause 4.17 of NSC/C requires the addition of one thirty-ninth to fluctuations and certain other costs so that the general contractor can deduct the discount without reducing the original amounts.

It hardly needs saying that if the payment to a sub-contractor is not made by the general contractor at all but direct by the Employer for any reason, then the former is not entitled to cash discount or a sum equivalent to it.

Statutory undertakings (e.g. electricity and gas boards) do not normally give cash discounts. This is recognized in the JCT Form in clause 6.3 by the express exclusion in such cases of the provisions of clause 35, where the work is executed by the statutory undertaker 'solely in pursuance of its statutory obligations'. The fact that the p.c. sums for work or services by statutory undertakers are net should be made clear in the tender documents. The tenderer is usually told that he will be deemed to have made allowance in the profit addition for any extra amount he may require in lieu of cash discount. In consequence, there will be no need for any adjustment to be made to the net totals of invoices in the manner illustrated above.

It follows from the wording of clauses 6.2 and 6.3 that, where a local authority or statutory undertaker is nominated to carry out works outside the scope of its statutory obligations (e.g. electrical and gas installations in buildings), then it is a nominated sub-contractor and the provisions of clause 35 apply, including the discount provisions.

General contractor's profit

Prime cost sums are deemed to be exclusive of the general contractor's profit. Provision is made in bills of quantities and specifications for a separate addition which is normally calculated as a percentage of the p.c. sum, the tenderer stating the percentage that he requires. When the contract sum is adjusted by the omission of p.c. sums and the substitution of actual costs incurred, the profit amounts will also be adjusted *pro rata*, by applying the same percentages to the actual costs as those inserted in the tender document against the respective p.c. sums.

Sometimes the profit additions may be shown as lump sums, without any indication of the method of calculation. In those cases also, such sums should be adjusted *pro rata* to the actual costs.

In the event of the Employer paying a sub-contractor direct for any reason, the general contractor is still entitled to the profit addition, as this is a management fee for arranging, organizing and taking responsibility for the sub-contract or for arranging and checking the supply of goods and materials, as the case may be.

Attendances

A general contractor has to 'attend upon' nominated sub-contractors, as he is responsible for their work and for the integration of the sub-contract works into the main contract works. So, in his own interests, he needs to provide facilities to sub-contractors so that their work may proceed unimpeded.

Tenderers are directed in the tender documents to allow for all such attendances in their tender sum.

When there are bills of quantities, 'general attendances' on all nominated sub-contractors are given together in one item; individual items are given for 'special attendances' on each sub-contractor as required. The former are defined in the *Standard Method of Measurement*, clause A42, as 'including the use of Contractor's temporary roads, pavings and paths, standing scaffolding ... standing power-operated hoisting plant, the provision of temporary lighting and water supplies, clearing away rubbish, provision of space for the sub-contractor's own offices and the storage of his plant and materials and the use of messrooms, sanitary accommodation and welfare facilities provided by the contractor for his own use'.

'Special attendances', defined in clause A51 of the *Standard Method of Measurement*, include such facilities as scaffolding, access roads and hardstanding additional to that which would otherwise be required, unloading and positioning heavy and large items, storage accommodation and provision of power supplies.

The prices included in tenders for attendances are usually treated as fixed sums which, unlike the profit additions, are not normally adjustable *pro rata* to the actual cost of the sub-contract works. This is because the kind and amount of work or services involved in the attendances remain the same, whether the actual costs are higher or lower than the p.c. sum. Only in the event of the sub-contract works being varied from what was originally envisaged to the extent of affecting the nature or the scope of the attendances, would the attendance amounts be varied.

Sometimes contractors show prices of attendance items as percentages of the p.c. sums in the same way as the profit items. In such cases, the amounts should still be treated as lump sums and the surveyor would be wise to obtain the contractor's agreement to this and to the deletion of the percentage rates, prior to the signing of the contract.

Examples of adjustment of p.c. and provisional sums

The following examples relate to the contract of which particulars are given in Appendix A. The list of p.c. and provisional sums included in the bills of quantities is given on pp. 337 and 338.

During the course of the contract, the architect issued Instructions nominating the sub-contractors and requiring orders to be placed with suppliers, as indicated below.

Nominated sub-contractors

Piling

The tender dated 12 January 1994, submitted by Drivas & Co. Ltd, in the sum of £11,340 was accepted in AI No. 1 dated 24 February 1994. Drivas's account dated 26 June 1994 was for the same as quoted, i.e. £11,340.

Asphalt tanking and roofing

The tender dated 3 February 1994, submitted by Asphaltank Ltd, in the sum of £13,725 was also accepted in AI No. 1 dated 24 February 1994. Asphaltank's account dated 3 May 1995 totalled £14,107, made up of £13,725 as quoted, £58 for increased costs of materials, £117 for dayworks resulting from AI No. 10 for alterations to work already executed and £207 for labour costs increases (made up of £201.83 plus one thirty-ninth). The £58 was deducted by the surveyor as inadmissible on the ground that asphalt was not on the list of basic prices of materials attached to the tender (see Chapter 9). The amount set against the p.c. sum, therefore, was £14,049.

Electrical installation

The tender dated 10 February 1994, submitted by Sparkes & Co. Ltd, in the sum of £18,000 was accepted and the sub-contractor was nominated in AI No. 1 dated 24 February 1994. Sparkes' account was dated 3 October 1995 and totalled £18,402.30, the extra £402.30 on the quoted amount being for net labour costs increases plus one thirty-ninth. This total was accepted and set against the p.c. sum.

Lifts installation

The tender dated 26 May 1994, submitted by Wayward Lifts Ltd, in the sum of £33,975 was accepted and the sub-contractor nominated in AI No. 21 dated 3 July 1994. Wayward Lifts submitted an account dated 20 December 1995, which totalled £35,804.79, which included £1152.36 for increases in the costs of steel and £677.43 for labour costs increases (both sums including the addition of one thirty-ninth to the net amounts). These amounts were accepted, steel being on the list of basic prices of materials attached to the tender.

All the above-mentioned quotations, tenders and accounts showed the correct cash discount.

Nominated suppliers

Kitchen fitments

A quotation dated 9 November 1994 was given by G. F. Joinery Ltd, amounting to £3215. In AI No. 40 dated 16 November 1994, the architect instructed the contractor to place an order in accordance with the quotation. The contractor telephoned the architect on 20 November pointing out that the quotation allowed only 2½ per cent cash discount and the AI was not therefore in accordance with the contract. The architect agreed but said that as there was a delay in delivery of two months at present, it would help if the contractor would place the order and the discount would be allowed for in the settlement of accounts. The contractor agreed to this and placed the order. Delivery of the fitments began in mid-January and was completed in early August. An invoice dated 25 August 1995 was rendered by G. F. Joinery but was endorsed 'net monthly account'. When the invoice was passed to the surveyor, he amended it to read '2½ per cent monthly account' to agree with the quotation and then adjusted the total to £3299.60 as shown in Figure 7.1, to allow for the correct discount.

Ironmongery

NKG Ltd submitted a quotation dated 13 October 1994 for £2923 and the contractor placed an order accepting it on 9 November 1994, as instructed in AI No. 38 dated 6 November. An account dated 1 May 1995 was rendered which was in accordance with the quotation. A second account dated 4 July 1995, for additional

Adjustment of prime cost and provisional sums*

Summary

	Omissions £ p	Additions £ p
Piling	378 00	
Asphalt roofing and tanking	2 258 55	
Electrical installation		422 42
Lifts installation		1 685 03
Kitchen fitments	315 42	
Ironmongery		371 36
Sanitary appliances	391 23	
Water main connection		12 29
Electricity mains connection		21 60
Gas main connection		10 40
Sewer connections		23 81
Dayworks	4 050 00	
Employer's telephone call charges	262 43	
Security installation		231 25
External light fittings	1 558 00	
Landscaping		341 13
Contingencies	8 000 00	
	17 213 63	3 119 29
Less Additions	3 119 29	
NET OMISSIONS CARRIED TO FINAL ACCOUNT SUMMARY	14 094 34	

* The adjustment of provisional sums, if numerous, may be shown
  separately from p.c. sums.

Figure 7.1 Adjustment of prime cost and provisional sums

Details of the adjustments summarized above:

		Omissions		Additions	
		£	p	£	p
Piling					
P.c. sum as item 144A		11 700	00		
Profit as item 144B	5%	585	00		
Drivas & Co. Ltd's account dated					
26/6/94				11 340	00
Profit *pro rata* item 144B				5% 567	00
		12 285	00	11 907	00
Less Additions		11 907	00		
Net omission carried to Summary		378	00		
Asphalt tanking and roofing					
P.c. sum as item 144E		16 200	00		
Profit as item 144F	5%	810	00		
Asphaltank Ltd's account dated 3/5/95				14 049	00
Profit *pro rata* item 144F				5% 702	45
		17 010	00	14 751	45
Less Additions		14 751	45		
Net omission carried to Summary		2 258	55		
Electrical installation					
P.c. sum as item 145A		18 000	00		
Profit as item 145B	5%	900	00		
Sparkes & Co. Ltd's account dated 3/10/95				18 402	30
Profit *pro rata* item 145B				5% 920	12
		18 900	00	19 322	42
Less Omissions				18 900	00
Net addition carried to Summary				422	42

Figure 7.1 continued

		Omissions £ p	Additions £ p
Lifts			
P.c. sum as item 145E		34 200 00	
Profit as item 145F	5%	1 710 00	
Wayward Lifts Ltd's account dated 20/12/95			35 804 79
Profit *pro rata* item 145F	5%		1 790 24
		35 910 00	37 595 03
Less Omissions			35 910 00
Net addition carried to Summary			1 685 03

		Omissions £ p	Additions £ p
Kitchen fitments			
P.c. sum as item 146A		3 600 00	
Profit as item 146B	5%	180 00	
G. F. Joinery Ltd's invoice dated 25/8/95			3 215 00
Less incorrect cash discount	2½%		80 38
			3 134 62
Add correct cash discount	¹⁄₁₉		164 98
			3 299 60
Profit *pro rata* item 146B	5%		164 98
		3 780 00	3 464 58
Less Additions		3 464 58	
Net omission carried to Summary		315 42	

Figure 7.1 continued

		Omissions		Additions	
		£	p	£	p
Ironmongery					
P.c. sum as item 146C		2 700	00		
Profit as item 146D	5%	135	00		
NKG Ltd's invoice dated 1/5/95				2 923	00
NKG Ltd's invoice dated 4/7/95				130	68
				3 053	68
Profit *pro rata* item 146D			5%	152	68
		2 835	00	3 206	36
Less Omissions				2 835	00
Net addition carried to Summary				371	36

		Omissions		Additions	
Sanitary appliances					
P.c. sum as item 146E		4 320	00		
Profit as item 146F	5%	216	00		
Saniware Ltd's invoice dated 11/5/95				3 947	40
Profit *pro rata* item 146F			5%	197	37
		4 536	00	4 144	77
Less Additions		4 144	77		
Net omission carried to Summary		391	23		

		Omissions		Additions	
Water main connection					
Provisional sum as item 146G		360	00		
Thames Water Board's account					
dated 21/9/94				354	56
Profit			5%	17	73
		360	00	372	29
Less Omissions				360	00
Net addition carried to Summary				12	29

Figure 7.1 continued

	Omissions £ p		Additions £ p
Electricity mains connection			
Provisional sum as item 147A	360 00		
South Eastern Electricity Board's account dated 13/10/94			363 43
Profit		5%	18 17
	360 00		381 60
Less Omissions			360 00
Net addition carried to Summary			21 60

	Omissions £ p		Additions £ p
Gas main connection			
Provisional sum as item 147B	540 00		
South Eastern Gas Board's account dated 2/11/94			524 19
Profit		5%	26 21
	540 00		550 40
Less Omissions			540 00
Net addition carried to Summary			10 40

	Omissions £ p		Additions £ p
Sewer connections			
Provisional sum as item 147C	900 00		
Urbiston Borough Council's account dated 2/10/94			879 82
Profit		5%	43 99
	900 00		923 81
Less Omissions			900 00
Net addition carried to Summary			23 81

Figure 7.1 continued

| | Omissions | | Additions | |
	£	p	£	p

Other work

Dayworks

| Provisional sum as item 147D | 4 050 00 | |
| Net omission carried to Summary | 4 050 00 | |

Employer's telephone call charges

Provisional sum as item 147D	750 00	
British Telecom accounts dated		
10/6/94		185 26
9/9/94		42 15
11/12/94		40 32
9/3/95		58 76
13/6/95		52 40
7/9/95		49 34
14/12/95		36 12
		464 35
Profit	5%	23 22
	750 00	487 57
Less Additions	487 57	
Net omission carried to Summary	262 43	

Security installation

Provisional sum as item 147E	2 000 00	
Smith Security Alarms Ltd's account		
dated 7/9/95		2 125 00
Profit	5%	106 25
	2 000 00	2 231 25
Less Omissions		2 000 00
Net addition carried to Summary		231 25

Figure 7.1 continued

	Omissions £ p	Additions £ p
Landscaping		
Provisional sum as item 147F	2 150 00	
Surrey Landscapes Ltd's account dated 30/11/95		2 372 50
Profit 5%		118 63
	2 150 00	2 491 13
Less Omissions		2 150 00
Net addition carried to Summary		341 13

	Omissions £ p	Additions £ p
Contingencies		
Provisional sum as item 147G	8 000 00	
Net omission carried to Summary	8 000 00	

Figure 7.1 continued

ironmongery ordered on 11 April arising from AI No. 56, totalled £130.68. Both accounts were in order and included the correct discount.

Sanitary appliances

Saniware Ltd quoted on 3 July 1994 for sanitary appliances for the sum of £3924.90. This was accepted on 27 July 1994, as AI No. 25 dated 21 July. Saniware Ltd's invoice dated 11 May 1995 totalled £3947.40, made up of the sum quoted plus £22.50 for special packings, and this was accepted by the surveyor.

Statutory bodies

The architect issued AI No. 2 on 3 March 1994 instructing the contractor to place orders with the Water, Electricity and Gas Boards for mains connections and with Urbiston Borough Council for sewer connections. The work was carried out and, in due course, accounts were received as follows, each account demanding payment in full, and being marked 'Net'.

Water main connection

Thames Water Board's account dated 21 September 1994 was for £354.56.

Electricity mains connection

South Eastern Electricity Board's account dated 13 October 1994 was for £363.43.

Gas main connection

South Eastern Gas Board's account was dated 2 November 1994 and was for £524.19.

Sewer connections

The Urbiston Borough Council's account was dated 2 October 1994 and was for £879.82.

Prime cost and provisional sums

Employer's telephone call charges

Accounts were received from British Telecom at quarterly intervals between June 1994 and December 1995, the first of which included the installation charge.

Security installation

The contractor was instructed in AI No. 80, dated 15 August 1995, to place an order with Smith Security Systems Ltd for an entry security system in accordance with their quotation dated 11 July 1995 amounting to £2125.00. Their account dated 7 September 1995 was for that sum. Profit was added at the same rate (5 per cent) as was shown in the bills for other nominated sub-contracts.

External light fittings

This work was omitted from the contract in AI No. 89, dated 3 November 1995.

Landscaping

The architect instructed the contractor in AI No. 84 dated 1 September 1995 to place an order with Surrey Landscapes Ltd for landscape work in accordance with their quotation dated 26 August 1995 in the sum of £2372.50. Their account dated 30 November 1995 was for that amount. Profit was added at the same rate (5 per cent) as was included in the bills for other nominated sub-contracts.

Figure 7.1 shows a method of presenting the adjustment of p.c. and provisional sums in accordance with the foregoing data. An alternative format is preferred by some surveyors, which differs from that given in the example in that the total omission and total addition relative to each p.c. sum are carried over to the Summary and then the net omission or net addition is calculated from the totals. Figure 7.2 shows what the Summary looks like when using this alternative format. The first alternative has the advantage that the net saving or extra resulting from the adjustment of each p.c. sum is shown. If this additional information is not of any particular use, then it does not matter which method is used.

General contractor carrying out works covered by p.c. sums

A general contractor may wish to carry out work himself for which a p.c. sum has been included in the contract documents or which has arisen under an AI. The JCT Form provides for this in clause 35.2, but with limitations. The limitations imposed by the JCT Form are:

(*a*) the work must be such as the general contractor does directly in the normal course of his business, although if he is allowed to do the work he may sub-let it, provided the architect agrees;

Adjustment of prime cost and provisional sums*

Summary

	Omissions	Additions
	£ p	£ p
Piling	12 285 00	11 907 00
Asphalt roofing and tanking	17 010 00	14 751 45
Electrical installation	18 900 00	19 322 42
Lifts installation	35 910 00	37 595 03
Kitchen fitments	3 780 00	3 464 58
Ironmongery	2 835 00	3 206 36
Sanitary appliances	4 536 00	4 144 77
Water main connection	360 00	381 60
Electricity mains connection	360 00	372 29
Gas main connection	540 00	550 40
Sewer connections	900 00	923 81
Dayworks	4 050 00	–
Employer's telephone call charges	750 00	487 57
Security installation	2 000 00	2 231 25
External light fittings	1 558 00	–
Landscaping	2 150 00	2 491 13
Contingencies	8 000 00	–
	115 924 00	101 829 66
Less Additions	101 829 66	
NET OMISSIONS CARRIED TO FINAL ACCOUNT SUMMARY	14 094 34	

* The adjustment of provisional sums, if numerous, may be shown
separately from p.c. sums.

Figure 7.2 Adjustment of prime cost and provisional sums: summary
by alternative format

(*b*) the contractor must have given notice, prior to the drawing-
up of the contract, that he wishes to carry out such work,
indicating what the items of work are, so that they may be set
out in the Appendix to the Conditions. (If a p.c. sum has
arisen because of an AI, then the notice should be given as
soon as the AI is received.);

(c) the architect must be willing to accept a tender from the contractor for such work; if not, the contractor will not be permitted to do the work himself. Such a refusal will usually be due to the architect preferring a specialist sub-contractor to do it, in whom he has more confidence to provide a satisfactory standard of work.

When the general contractor is permitted to submit a tender and his tender is accepted, the question then arises as to whether he is entitled to be paid the profit addition included in the contract sum. The answer appears to be that it depends on whether the tender for the sub-contract works was submitted in competition or not. If it were, then the general contractor is entitled to the profit addition just as if a sub-contractor were to do the work. If it were not in competition, then the contractor's price should include the full amount of profit, so that, in effect, the p.c. sum is converted into a provisional sum and the contractor should be asked to include in his tender the total amount of profit required.

Another question relates to 'special attendances' (see p. 89). They may involve extra expense such as providing special scaffolding or staging or special plant. If the general contractor tenders in competition with sub-contractors, he would be at a dis-advantage if he had to include such extra costs, whereas they would not do so. In those circumstances, the amounts in the contract sum for the relevant items of 'special attendances' would stand. If the tender is not in competition, the contractor should be instructed to include for all 'attendances' in his tender and any attendance amounts included in the contract sum should be deducted.

A situation which may arise, involving a desire by the general contractor to do sub-contract work himself, is where a nominated sub-contractor defaults before completion of his sub-contract. The general contractor may want to avoid the delay which would be occasioned by the need for the architect to re-nominate, so that the rest of the work may not be held up.

The JCT Form, 1963 Edition, made no direct provision for dealing with this situation. It is not surprising therefore, that this has been the subject of legal action. The classic case is *North-West Metropolitan Hospital Board v T. A. Bickerton Ltd (1970) 1 All ER 1039*. The House of Lords expressed the view in this case that the general contractor did not have the right (nor the duty) to do work for which p.c. sums had been included in the contract documents.

This appears to conflict with the express provisions of clause 35.2 discussed above but, presumably, this (or more correctly, the corresponding clause 27 (g) in the 1963 Edition, then in force) was outside the scope of the circumstances of the case being considered by the House of Lords and thus was excepted from their Lordships' expressed view. It was also held that, if a nominated sub-contractor defaults, a re-nomination by the architect must follow.

The particular point for the surveyor to note is their Lordships' view that the total cost of the sub-contract works in such circumstances is the responsibility of the Employer, even though it may have been increased due to a new sub-contractor requiring to be paid more to finish the work than it would have cost had the original sub-contractor completed. The case has been discussed at length elsewhere.[3]

The 1980 Edition of the JCT Form, however, does deal with the matter of re-nomination by the architect. The provisions of clause 35.24 were no doubt introduced as a consequence of the *Bickerton* judgment. Briefly, the clause requires the architect to make a further nomination of a sub-contractor to finish the sub-contract works whenever a nominated sub-contractor fails to complete. Any additional expense arising as a consequence is the responsibility of the Employer, unless the sub-contractor has determined his sub-contract because of default by the contractor, in which case the additional expense is recoverable by the Employer from the contractor.

In the case of *Fairclough Building Ltd v Rhuddlan Borough Council (1983) QB 3 ConLR 38*, the decision in the *Bickerton* case was upheld. The subsequent appeal was dismissed by the Court of Appeal in 1985. In the judgment, it was stated that the new sub-contract must include for remedying all defects in the earlier sub-contract work, since the main contractor was not liable to carry out such remedial work. If such work were not included, the main contractor could refuse to accept the re-nomination.[4]

Provisional sums

Provisional sums are usually intended to meet the cost of work not determined or defined at the time of tendering, but which normally will be carried out by the general contractor. In due course, the architect will issue, in accordance with clause 13.3 of the JCT Form, written instructions to the contractor from time to time as to what work he is required to do, which will be paid for out of the Provisional sums.

When the contract sum is adjusted at the end of the contract, the Provisional sums, like p.c. sums, will be deducted and the total value of the work actually carried out will be substituted. In order to ascertain what that total cost is, the work will (if carried out by the general contractor) be measured and valued as executed and will be priced in the same way as varied work is priced (see Chapter 8). Such total cost may also include work valued as daywork. Alternatively, the contractor may be asked to submit a tender for carrying out the work and, if his tender is accepted, the amount tendered will be set against the appropriate Provisional sum in the settlement of the accounts.

As the work covered by Provisional sums will often be carried out by the general contractor and not by a sub-contractor, no items of profit or attendances are attached to such sums in the bills of quantities. In that case, the valuation of the work carried out is inclusive of profit.

However, the work for which a Provisional sum has been provided may, if the architect so decides, be carried out not by the general contractor but by a sub-contractor nominated by the architect. Then, the sub-contractor's account will be dealt with in the same way as in the case of a p.c. sum. An item of profit will be added at the percentage rate used for the profit items on p.c. sums for sub-contractors' work in the bills and attendance items, as appropriate, will also be added. Thus, the total of the sub-contractor's account (as submitted or as amended) plus profit and attendance amounts, will be set against the Provisional sum in the final account.

Again, the architect may wish to nominate a particular supplier to supply materials or goods which are to form part of the work for which a Provisional sum has been included in the contract sum. In this case, the contractor will be entitled to a profit addition on the value of the nominated supplier's invoice at the percentage rate of the profit items on comparable p.c. sums in the bills. The remainder of the work will be measured and valued in the same way as varied work, including items for fixing the goods supplied by the nominated supplier.

It should be noted that the situations described in the last two paragraphs are envisaged in clauses 13.3, 35.1.2 and 36.1.1.2 of the JCT Form. Examples of the adjustment of provisional sums are included in Figure 7.1.

The question has been raised[5] as to the correct procedure where the actual value of work which is the subject of a Provisional sum exceeds the amount of that sum. In practice, that situation is usually dealt with simply by a net addition to the contract sum of the excess amount. Where the excess is large, however, a contractor might be disposed to claim an extension of time under clause 25 and reimbursement of any additional loss or expense under clause 13.5.5 on the grounds put forward in the article referred to.

References

1. *Standard Method of Measurement of Building Works, Seventh Edition* (London: RICS Books, 1988), p. 14.
2. POWELL-SMITH, Vincent, *Contractors' Guide to the JCT's Standard Forms of Building Contract* (Sutton: IPC Building & Contract Journals Ltd, 1977), p. 24.
3. ibid., p. 24.
4. POWELL-SMITH, Vincent, *A Contractor's Guide to the JCT Standard Form of Building Contract (JCT80), 2nd edn* (London: Legal Studies and Services (Publishing) Ltd, 1988).
5. SIMS, J., Sims on Contract Practice, *Building*, 1974, Vol. 227, No. 49, p. 98.

8

Variations

Definition and origin of variations

The definition of the term 'variation' in clause 13.1 of the JCT Form indicates the wide scope for the exercise of the architect's power to vary the Works.

Variations may arise in any of the following situations (references are to the JCT Form):

(a) when the architect needs or wishes to vary the design or the specification;
(b) when a discrepancy is discovered between any two or more of the contract documents (*see* clause 2.3);
(c) when a discrepancy is discovered between any statutory requirement and any of the contract documents (*see* clause 6.1);
(d) when an error in or omission from the contract bills is discovered (*see* clause 2.2);
(e) when the description of a provisional sum for defined work (*see* page 80) in the contract bills does not provide the information required by General Rule 10.3 of SMM7 (*see* clause 2.2).

By far the largest number of variations arises under (a). In cases (a), (b) and (c), the architect is required to issue instructions but that is not necessary in cases (d) and (e). All such instructions must be in writing. Consequently, although the architect may order any variation that he wishes, the contractor is bound to carry out such an order only if it is in the form of a written instruction issued by the architect (unless the contractor makes a reasonable objection in writing to him under clause 4.1.1). If the contractor carries out any variation involving him in additional expense, and the variation has not been the subject of a written Architect's Instruction (AI), he runs the risk of being unable to recover the extra costs.

The contractor is safeguarded, however, in the event of instructions being given orally. Under clause 4.3 of the JCT Form, an oral instruction 'shall be confirmed in writing by the contractor to the architect within seven days'. The architect has the opportunity during seven days following receipt of that confirmation to dissent, otherwise the instruction becomes effective immediately. Failing such action, the architect may issue a confirmation at any later time but prior to the issue of the Final Certificate. Where oral instructions have allegedly been given and complied with but not subsequently confirmed, some surveyors ask the architect to issue confirmation, if he is willing to do so.

The architect may thus issue a written instruction which sanctions a variation already carried out in compliance with oral instructions. Clause 13.2 of the JCT Form goes further, enabling the architect to sanction subsequently any variation which a contractor has already carried out not in pursuance of an AI, oral or written.

On contracts where there is a clerk of works, any directions which he may properly give to the contractor (whether orally or in writing) become effective only if confirmed in writing by the architect within two working days (clause 12).

It is necessary, therefore, for the surveyor to satisfy himself, before giving financial effect to any variation, that written instructions have been given by the architect, ordering or sanctioning the variation, or if the contractor claims to have confirmed oral instructions, that the architect does not dissent from such confirmation. The surveyor has no authority to adjust the Contract Sum in respect of a variation which has not been put into the form of a written AI or been otherwise confirmed in writing (*see* comment on p. 5).

It should be appreciated that all Architect's Instructions do not necessarily constitute or contain variations to the contract. Some may be explanatory, others may direct how provisional sums and p.c. sums are to be expended, yet others may instruct the contractor to remove from the site materials not in accordance with the contract. None of these AIs may constitute a variation to the Works.

The cause of a variation listed as (*d*) above arises under clause 2.2 of the JCT Form. This is where an error or omission has occurred in the compilation by the surveyor of the bills of quantities upon which a contract sum has been based. In such circumstances the error or omission must be rectified and 'shall be treated as if it were a variation'. It should be noted that this does *not* relate to errors made by the contractor when pricing bills of quantities (see p. 59).

Measuring variations

Few, if any, variations have no effect on the Contract Sum. In order, therefore, to adjust the Contract Sum adequately and satisfactorily within the Conditions of the contract, the quantity surveyor (and the contractor's surveyor, too, if he wishes) must ascertain the net extra cost or net saving involved. To do that, the affected work will have to be measured, except in those instances where the method of valuation makes it unnecessary, as discussed below. This means that the work originally designed, which will not be required as a consequence of the variation, must be identified, isolated and valued, and that which will be required to replace the omitted work must be measured and valued. Such measurement may be done from drawings or, if drawings are not available, by physically measuring the substituted work on the site after it has been carried out.

This same procedure will be necessary whether the contract is based upon bills of firm quantities or on specification and drawings only. In the case of contracts based upon bills of approximate quantities and those based on a schedule of rates, the whole of the Works will have to be measured as described above. In all cases the valuing will be done in the manner discussed in the section below headed 'Valuing variations'.

It should be appreciated that in the case of 'without quantities' contracts, the work omitted will have to be measured in detail, as well as the substituted work, there being no simple procedure of identifying bill items affected, as in the case of 'with quantities' contracts. Where there are bills, it will often be necessary to refer back to the original dimension sheets in order to identify exactly what was included in the bills for the work which has been varied. If the items affected constitute complete bill items, they can be identified by their item references either as a single item or a group of items, e.g. 'item 132B' or 'items 132B-G'. If a variety of items is involved, many of which are not complete bill items, the omissions can be identified by reference to the dimension column numbers, e.g. 'original dimensions, X-X cols. 95–103', the letter 'X' being written on the dimension sheets to indicate the commencement and end of the omitted work. The measurement of AIs Nos. 8 and 48 in the examples at the end of the chapter illustrates the alternative procedures.

The quantity surveyor has the duty laid upon him by clause 13.6 of the JCT Form, to provide the opportunity for the contractor to be present when measurements of variations are taken. Although the clauses do not specifically refer solely to measurements on site, in practice the provision is not normally applied to measurement from drawings done in the office. It does mean, however, that the surveyor must give the contractor's surveyor sufficient notice for him to be able to make arrangements to attend on site when measuring is to be done there.

When measuring on site, it should be remembered that it is intended by the Committee that prepared the Standard Method of Measurement that the rules of measurement in that document should apply equally to measurement on site and to measurement from drawings.[1] This is important, for example, when measuring excavation involving working space and earthwork support. The SMM requires these items to be measured, even though the surveyor doing site measuring may know that they have not actually been done. The dimensions of the excavation should be those which are minimum for the required construction, measured in accordance with the rules of the SMM, notwithstanding that the contractor may have excavated wider to avoid the use of earthwork supports or because the excavator bucket available was wider than needed. So, too, the lengths of timbers whose ends are not cut square (e.g. rafters) should be measured by their extreme lengths and inflexible floor tiles (e.g. quarry tiles) should be measured *between* skirtings, particularly where the latter are coved, and not between wall faces.

Measurements taken on site are normally entered in bound dimension books, which consist of pages with dimension column rulings similar to 'cut and shuffle' slips, or to traditional dimension paper but with only one set of columns instead of two. Alternatively, one may use normal dimension paper on a clip board which, although not so convenient to handle as a dimension book, does give the advantage of having all the measurements, whether taken on site or in the office, on the same size sheets.

Another way of dealing with both the measurements and the working-up involved, is to use A4 size paper ruled with dimension columns on the left and quantity and money columns on the right – i.e. similar to estimating paper. Thus, all the information – measurements, billing and pricing – is presented on the same sheets (see AIs Nos. 15, 23, 28 and 41 in the examples on pages 125–134).

Variation accounts

When the measurement has been done, the dimensions have to be worked up into variation accounts, which, in compliance with clause 30.6.1.2 of the JCT Form, should be completed and priced and a copy supplied to the contractor not later than 9 months after Practical Completion. In practice, this is seldom achieved. Nevertheless, the surveyor should do his best to complete the variation accounts by the expiry of the period stated in the contract as the Employer is in breach of contract if he does not do so.

The format in which the variation accounts are drawn up varies between one firm of surveyors and another. Some firms bill omissions and additions under each AI separately. Others combine all the omissions together and all the additions together, each group being classified under appropriate Work Section headings. The merit of the former method is that the cost of each variation is apparent, together with separate totals of omissions and additions. The main disadvantage is that the variation accounts are almost certain to form a thicker document due to the repetition of items in different variations and also because of the extra space taken up with separate totals. The second method results in less items and a more compact document which takes less time to prepare and price. If all that is required is the amount of the net extra or saving, then the latter format would seem to be adequate. If details of the costs of the individual variations is needed or desired, then the former method of presentation will probably be necessary. Apart from that, the format may be dictated by custom in individual surveying organizations.

Valuing variations

There are several ways of valuing variations, the choice in a particular case being that which is appropriate to the circumstances. They are:

(a) by the inclusion in the variation accounts of a lump sum in accordance with a quotation submitted by the contractor and accepted by the architect;
(b) by pricing measured items in the variation accounts;
(c) by ascertaining the total prime cost of additional work and applying appropriate percentage additions.

It should be noted that clause 13.4.1.1 of the JCT Form lays the duty of valuing variations upon the quantity surveyor, who has the right to decide which is the appropriate method and means of valuing them in each situation. Therefore, he is not bound to accept a statement on an AI to the effect that a variation is 'to be carried out on daywork' or 'to be paid as daywork'. In any case, such a statement may be contrary to the provisions of the JCT Form.

The alternative valuation methods listed above merit further discussion as follows.

Contractor's accepted quotation

Where the architect provides adequate information in an AI, clause 13A of the JCT Form makes provision for the contractor (if he so agrees), to submit to the surveyor a quotation (known as a '13A Quotation') for carrying out the specified work. Such work may include work to be carried out by nominated sub-contractors from whom the contractor may obtain quotations (known as '3.3A Quotations') for inclusion in his 13A Quotation.

The 13A Quotation must show separately and in sufficient detail the value of the adjustment to the Contract Sum, including the effect on any other work (e.g. omitted work and adjustment of preliminary items) with supporting calculations based on rates and prices in the Contract Bills as appropriate; any adjustment of time required for completion of the Works; the amount to be paid in lieu of the value of direct loss and/or expense under clause 26.1; and a fair and reasonable sum for preparing the 13A Quotation.

The Employer has seven days from receipt by the surveyor of the 13A Quotation in which to accept it in writing. If so accepted, the architect must immediately confirm the acceptance in writing to the contractor, stating the consequential adjustment to the Contract Sum and a revised Completion Date, and authorizing the contractor to accept any 3.3A Quotations included in the 13A Quotation. The Contract Sum will be adjusted accordingly in the Final Account.

This is an attractive way of valuing a variation because it eliminates the need for time-consuming measurement and pricing. However, it contains the conditions for inflated prices: no competition, and the liability that the quotation will be insufficiently scrutinized due to pressure of time and also the architect's understandable anxiety to avoid delay. But provided the surveyor makes an approximate estimate of *net* extra cost (i.e. taking account of omissions as well as additions), and the contractor's price bears a reasonable relationship to the estimate, then the danger will be minimized.

If the contractor does not agree to provide a 13A Quotation when so requested in an AI, the variation will be valued in the normal way under the valuation rules in clause 13.5. If the Employer has not accepted a 13A Quotation by the end of the seven days allowed for acceptance, the architect must instruct the contractor either that the variation is to be carried out and valued under the valuation rules in clause 13.5 or that it is not to be carried out. In either case, a 'fair and reasonable amount' is to be paid to the contractor for preparing the 13A Quotation.

It should be noted that clause 13A does not place any limit on the scope or the value of the work included in the variation which is the subject of the 13A Quotation. It follows that the architect can, within the Conditions of Contract, vary the works to include major changes in the scope and value of the Contract.

Pricing measured items

The JCT Form in clause 13 is explicit about the way measured items (omissions as well as additions) should be priced. The prices (in particular, unit rates) in the bills or schedule of rates, as the case may be, are to be used. This is subject to the proviso that the character of the work and the conditions of its execution are similar to that in the project as originally envisaged. This covers the pricing of all (or most) omissions and much of the additions (although some omission items in 'without quantities' contracts may not coincide with items in the submitted schedule of rates).

Other items will bear no relationship or comparison at all with items in the contract bills, or may not be carried out under similar conditions to apparently similar items in the bills. In these cases, a 'fair valuation' is to be made. This means that a unit rate may have to be built up from the prime cost of the necessary materials and using labour constants valued at 'all-in' labour rates and with any appropriate allowances for plant and with additions for overheads and profit. If an item is a common one or of small total value, the surveyor's knowledge of prices will usually enable him to put a fair price to it without the necessity for a detailed synthesis on the lines just described.

In between the two foregoing groups of items, there is a third group in which the items are not exactly the same as but bear a fairly close resemblance to items in the bills or schedule of rates. These should be priced *on the basis of* the prices of the comparable items. That means in effect, that the pricing of such items is to be done, as far as possible, at the same general level of prices as is contained in the contract documents. Again, where an item of work is common or relatively inexpensive, the surveyor will usually be able to 'assess' a comparable rate to the similar bill or schedule item. Where the item is an expensive one in total cost terms, or where there is any dispute, it may be necessary to analyse the contract rate and then to synthesize the new rate on the basis of the analysis. In any case, the student should be able to do price analysis and synthesis and the examples at the end of the chapter will illustrate the process.

Perhaps it should be emphasized that, in practice, it is only necessary for relatively few items in variation accounts to be priced by a process of analysis and synthesis. It should be realized, however, that, in theory, the same process is being carried out when a comparable rate is 'assessed'. As in many fields, experience enables one to short-circuit longer procedures, but this can only be done safely once the underlying principles and procedures have been fully understood and appreciated. Then, (to give a simple illustration) the student will know that a 100 mm thick concrete slab should not be priced at half the price of a 200 mm one to the same specification, even without the experience to be able immediately to suggest a reasonable price for the thinner slab.

Prime cost plus percentage additions

This method of valuing variations is commonly known as *daywork* and bears a close resemblance to the valuation of work under the prime cost or cost reimbursement type of contract (see p. 17). Daywork is a method of payment by the reimbursement to the contractor of the prime cost of all materials, labour and plant used in the carrying out of the work, with a percentage addition to the total cost of each of those groups for overheads and profit.

Daywork is subject to the same objection as that levied against prime cost contracts, namely, that its use generally results in higher costs to the Employer than when a 'measure and value' basis is used. Not surprisingly, therefore, daywork is favoured by contractors and deprecated by the Employer's professional advisers.

This generally acknowledged characteristic of daywork is the reason for the restriction of its use in the Standard Forms to situations where 'work cannot properly be valued by measurement'. This provision should be strictly observed in the Employer's interest. If work is capable of adequate measurement, it can usually also be priced using unit rates determined as described under *Pricing measured items* above.

Such measurement must be capable of being 'properly' done, which would exclude, for example, measuring some alteration or adaptation works by a single 'number item' with a long and complicated description, merely to avoid paying the contractor daywork. In any case, to price such an item would probably involve the surveyor in asking the contractor how long had been spent on the work, what materials and how much of them had been used, and what plant – which comes down to something very much like daywork in the end! The surveyor should recognize and accept that the contractor has the right to be paid on a daywork basis where appropriate and that the Employer has the right to pay on a measured basis where that can reasonably be done.

As with prime cost contracts, it is necessary to have a precise definition of what is included in the prime cost in each of its three divisions – materials, labour, and plant. The JCT Form refers to the document *'Definition of Prime Cost of Daywork carried out under a Building Contract'*[2] as the basis to be used for the calculation of the prime cost, or, in the case of specialist work, a comparable document appropriate to that work, if one exists.

The respective percentages to be added to the totals of each section of the prime cost will be those included in the contract document, i.e. bills of quantities or schedule of rates. If by some oversight, no percentages were included, then they would have to be agreed between the surveyor and the contractor's surveyor.

The contractor is required to submit 'vouchers' (commonly called 'daywork sheets') not later than the end of the week following that in which the daywork was completed, giving full details of labour, materials and plant used in carrying it out. These vouchers should be verified by the architect or his authorized representative. If there is a clerk of works, he can usually be relied upon to do the verifying. The surveyor may be prepared to accept unverified daywork sheets where he is able to satisfy himself that the claim is reasonable, but to do so is not strictly within the terms of the Standard Forms. However, if unverified sheets are accepted and the hours and quantities shown on them are considered to be excessive, the surveyor should reduce them to reasonable amounts.

It should be noted that a signature denoting verification by the architect or his representative relates only to the materials used and the time spent[3] and not to the correctness or otherwise of any prices included. The proper valuation is for the surveyor to determine.

There are difficulties associated with verifying daywork vouchers but as they are problems for the architect rather than the quantity surveyor, it is not considered necessary to discuss them here; the reader is referred to a relevant article in *Building*.[4]

The rates and prices used for valuing daywork should be those current at the time the work was carried out, not those current at the date of tender. This is so that the application of the respective percentage additions for overheads and profit will be to the total costs. If the rates and prices current at the date of tender were used, the total payment for the daywork would be less (assuming some increases in costs since the tender date). The difference would not be made up by the reimbursement of such increases under the fluctuations provisions of the contract, even if such provisions were included. It is also because rates and prices current at the time the daywork was carried out are used for valuing it, that the value of such work is excluded from the provisions relating to fluctuations in clauses 38, 39 and 40 of the JCT Form.[5]

The rates of labour used in valuing daywork should be standard hourly base rates, irrespective of the actual rates paid. Where rates above the basic rates are paid, the excess comes out of the percentage addition. The method of computing standard hourly base rates for use in pricing dayworks is illustrated on p. 8 of the *Definition of Prime Cost of Daywork*.*

Plant costs may be calculated on the basis of the *Schedule of Basic Plant Charges* issued by the Building Cost Information Service (BCIS) and intended to be used solely for that purpose. If so, then the rates quoted will require updating. This may be done by a percentage addition before the appropriate percentage for overheads and profit is added, or by combining a suitable updating percentage with that for overheads and profit. Provision for both the use of the *Schedule* and for updating the rates ought to have been made in the contract bills or schedule of rates, as the case may be.

Figure 8.6 (p. 135) gives an example of a daywork sheet and Figure 8.5 (p. 133) shows the same information converted into a daywork account for incorporation into a variation account.

---

* For an updated version of the base rate computation, see BCIS *Guide to Daywork Rates*.[6]

Additional expense arising from variations

There is a possibility that the contractor may not always be adequately remunerated for what he is required to do in regard to variations under the terms of the Standard Forms. For example, a variation which increased the width of a number of windows might be ordered. The timing of the order might result in the contractor having bought lintels for the window openings which had been delivered before the variation was ordered, and which, in consequence, became surplus to requirements. The normal measurement of the omissions and additions involved would result in the omission of the shorter lintels and the addition of longer ones and no account would be taken of the surplus lintels. The possibility of additional expense arising from variations is recognized by the JCT Form in clause 13.5.7. The contractor should make a written application to the architect for reimbursement of the additional expense which he has incurred and upon receipt of such application if it is justified, the surveyor must ascertain the amount involved, which then must be added to the Contract Sum. Any such amount should also be included in the next interim certificate issued under clause 30.

Use of erroneous rates when pricing variations

Sometimes, a rate against a measured item in bills of quantities or in a schedule of rates is inadvertently underpriced or overpriced by the contractor and the error may have escaped the notice of the surveyor when examining the tender document before the contract was let, so that nothing was done about it (see p. 59). If, subsequently, the item is involved in a variation, then a difficulty may arise in the use of the erroneous rate, which must be resolved.

Suppose, for example, that an item of 'two-coat plaster on walls' had been priced at £0.69 per m^2 in mistake for £6.90 per m^2. No problem need arise over omissions, as the omitted quantity should, without doubt, be priced at the bill rate. But suppose the additions involve a much larger quantity than that omitted. At what rate should the addition quantity be priced? Is it inequitable to the contractor to price it at the underpriced rate? Is the Employer taking unfair advantage of a genuine error on the contractor's part?

The answers to these questions should be sought by asking 'what does the contract say?'. The JCT Form in clause 13.5.1 clearly states that bill rates shall be used, provided the work is of similar character and executed under similar conditions. There is no authority therefore for the surveyor to use a different rate, so long as the work is 'similar' in those respects. As in other aspects of contractual agreements, what has been freely agreed by both parties is binding upon them, notwithstanding that a term of a contract may subsequently prove to be disadvantageous to one of the parties. It seems clear, therefore, that the contractor must stand the loss (or be allowed to gain the benefit, as the case may be) of an erroneously-priced item that is involved in variations.

A compromise that has been suggested from time to time is that the erroneous rate should be applied only up to the limit of the quantity of the item in the contract bills to which it applies and that beyond that limit a fair rate should be substituted. This will appear to many people as an equitable solution although it must be said again that there is no direct authority in the JCT Form for so doing. The judgment in a case heard in the High Court in 1967, *Dudley Corporation v Parsons & Morris Ltd*, held that the total quantity of an item in a remeasurement should be priced at the erroneous rate in question, a decision which was upheld in the Court of Appeal in 1969. This appears to settle the matter.

The question has been raised whether, in these circumstances, the contractor can justifiably claim reimbursement of the loss resulting from applying the erroneous rate to the additions quantity, under the provision of clause 13.5.7 of the JCT Form. Legal opinion that such a loss could not be recovered under clause 11(6) of the 1963 Edition of the JCT Form[7] would seem to apply also under clause 13.5.7 of the 1980 Edition.

Adjustment of Preliminaries on account of variations

This matter has always been a cause for contention and disagreement. Some contractor's surveyors have argued that if the value of the job is increased by variations, then the Preliminaries should also be increased in proportion, on the grounds that, by and large, they are directly affected by the size and value of the contract.

On the other hand, some surveyors have contended that the Preliminaries should only be adjusted in very exceptional circumstances, i.e. when the size and scope of the job is significantly affected, as, for example, by the addition of another storey or another wing or block. Others have taken the intermediate position that if there has been an extension of time granted in respect of variations, then there may be grounds for adjusting part of the value of Preliminaries.

Editions of the JCT Form published before 1980 were of little or no help in providing a solution, as they made no specific reference to Preliminaries. The 1980 Edition, however, is specific on the matter in clause 13.5.3.3. This requires that, when valuing additional, omitted or substituted work, allowance is to be made for any addition to or reduction of preliminary items. It should be noted, however, that the provision relates only to work 'which can properly be valued by measurement' and so excludes other work, such as daywork. Also excluded is work pursuant to an AI for expenditure of a provisional sum for 'defined work'.

If the surveyor has to ascertain the value of any addition to or reduction of Preliminaries, how may he do so? He will first have to satisfy himself as to the proper total value of the Preliminaries. If there are bills of quantities and the Preliminaries are fully priced, the total value is apparent. If, however, they are not fully priced, an assessment of their full value will need to be made, because the JCT Form requires allowance to be made for additions to or reduction of the items. In order to make such allowance, the value of the items as originally intended needs to be known.

In the case of contracts on a 'without quantities' basis, the contract sum will need to be apportioned to its principal constituents, of which the Preliminaries form one, unless a Contract Sum Analysis has already been provided. When the total value of the Preliminaries has been determined, the surveyor must then ascertain the value of the individual items which comprise that part of the Contract Sum. Then he will be in a position to proceed with his valuation of any addition or reduction, as the case may be.

The pricing of Preliminary items by means of fixed charges and time-related charges was introduced into the *Standard*

Method of Measurement of Building Works for the first time in the Seventh Edition (SMM7). It is intended that in the preparation of bills of quantities, etc., opportunity should be given for a fixed charge and a time-related charge to be included for all items of Employer's requirements (clauses A30–A37) and contractor's general cost items (clauses A40–A44). It is considered that this facilitates the adjustment of Preliminaries where this is required by clause 13.5.3.3 of the JCT Form when valuing variations. The fixed-charge element for each item can be analysed further for valuation purposes, into sums expended early and those expended late in the contract, as indicated in Table 12.2, p. 183.

If an extension of time has been granted by the architect on account of variations, then those Preliminary items with a time-related content may be considered as ranking for a proportionate increase in that content, whilst the cost-related and fixed charge elements remain unchanged.

If no extension has been granted, there may still be justifiable grounds for adjusting the item prices. For example, a variation might result in plant being kept on site longer than otherwise envisaged or in additional supervision being required for the period during which the variation is being carried out. In such situations, the actual expense would need to be ascertained, which would require the assistance of the contractor in providing the necessary details.

It should be appreciated, however, that in no case should the total value of the Preliminaries be increased (or decreased) simply in proportion to the total value of variations. Any adjustment must be made with regard to the amount by which individual Preliminary items have been affected by variations.

Examples of variations

The following is a list of the first six variations contained in Architect's Instructions issued for the contract of which particulars are given in Appendix A. (AIs not included in the list were those giving instructions regarding the expenditure of p.c. sums and dealing with other matters which did not constitute variations.)

AI No.	*Subject of variation*
8	Willingdon Light Buff Facing Bricks in lieu of Hildon Tudor Brown Facing Bricks.
15	The roof slabs to the flats increased from 125 mm thick to 175 mm and the main reinforcement increased from 12 mm to 16 mm diameter
23	Living room windows in flats to be 1800 × 1500 mm in lieu of 1800 × 1200 mm.
28	Glazed screens to balconies to be glazed with pattern group 2 glass with putty in lieu of clear glass in beads.
41	Window removed from existing store, opening in wall altered as necessary, provision of door and frame complete with deadlock, handles and closer.
48	Omission of built-in bookcase units in living rooms.

Figures 8.1–8.6 show the taking-off and billing for the variations. The work was billed direct from the taking-off in the case of AIs Nos. 8 and 48. The other AIs illustrate the use of 'estimating paper' ruling as an alternative format.

The following is a commentary on the examples.

AI No. 8 Facing Bricks

This variation was necessitated by Hildon Brickworks Ltd ceasing the manufacture of Tudor Brown Facings. The *additions* items were all the same as the corresponding bill items apart from the change in bricks. As the latter were comparable in type and quality with those originally specified, the rates were adjusted merely to take account of the increased prime cost of the bricks, the value of the other materials and labour being unchanged. The price of the Hildon Tudor Brown Facings used by the contractor when pricing the bills was £211.80 per 1000 including delivery. The price of the Willingdon Light Buff Facings, delivered, was £227.04 per 1000 and this price had remained unchanged since before tender date. The rates for the 102 mm and 215 mm walls items were calculated as shown on p. 123.

A.I. No 8   £   p

Willingdon Light Buff facings in lieu
of Hildon Tudor Brown facings

OMISSIONS

		£	p
A	Items 46A – 47A inclusive	54 271	25
B	Items 144F – 145C "	11 978	35
	TOTAL OF OMISSIONS CARRIED TO SUMMARY	66 249	60

ADDITIONS

Facework in Willingdon Light Buff
facings and pointing in cement:
lime: sand (1:1:6) mortar as the
work proceeds.

					£	p
C	Walls, 102 mm thick, facework one side p.r. item 46A.	510	m²	30.71	15 662	10
D	Ditto, 215 mm thick, ditto, p.r. item 46B.	1272	m²	51.01	64 884	72
E	Ditto, 215 mm thick, facework both sides, p.r. item 46C.	12	m²	59.88	718	56
F	Extra over, facework to reveals, 102 mm wide, p.r. item 46D.	322	m	0.48	154	56
G	Ditto, ditto, 215 mm wide, p.r. item 46E.	16	m	0.48	7	68
H	Closing cavities, 50 mm wide, p.r item 46F.	42	m	3.66	153	72
J	Extra over, flush horizontal brick-on-end band, 215 mm wide, p.r. item 47A	218	m	11.11	2 421	98
K	Walls, 102 mm thick, facework both sides, p.r. item 144F	143	m²	39.58	5 659	94

carried forward   89 663   26

Figure 8.1

£ p

A. I No. 8 (cont).

brought forward   89 663   26

<u>Facework in Willingdon Light Buff facings (cont)</u>

Steps and Flats, Thames St., Swindon

		£ p	£ p
A	Walls, 215 mm thick, facework one side, p.r. item 144 G.	6 m² 51.01	306 06
B	Ditto, 215 mm thick, facework both sides, p.r. item 144 H.	46 m² 59.88	2 754 48
C	Extra over, facework to reveals, 102 mm wide, p.r. item 145A.	28 m 0.48	13 44
D	Ditto, ditto, 215 mm wide, p.r. item 145B.	11 m 0.48	5 28
E	Coping, brick-on-end, 215 mm wide, horizontal, p.r. item 145C.	182 m 11.54	2 100 28

TOTAL OF ADDITIONS CARRIED TO SUMMARY   94 842   80

Figure 8.1 continued

	£
Prime cost of Willingdon Light Buffs per 1000	227.04
Prime cost of Hildon Tudor Browns per 1000	211.80
Difference	15.24

102 mm walls
  61 bricks per m², Stretcher Bond, @
£15.24 per 1000	0.93
15% OHP*	0.14
	1.07
Bill rate, item 46A	29.64
Required rate	30.71

215 mm walls, facework one side
  81 bricks per m², Flemish Bond,
@ £15.24 per 1000	1.23
15% OHP	0.18
	1.41
Bill rate, item 46B	49.60
Required rate	51.01

* The contractor stated, upon request, that the tender included 15 per cent mark-up on net cost for overheads and profit.

AI No. 15 Roof slabs

This variation is shown using the format described on p. 110 in which the dimensions, billing and pricing are on the same sheet. The calculation of the *pro rata* rates for the additions was made as follows.

Concrete slab

The bill rate was first analysed to find the approximate rate used for the concrete at the mixer and this was then adjusted for the reduced labour input due to the increased thickness. It was considered that the increased size of reinforcing bars did not affect the placing of the concrete sufficiently to justify amendment of the labour content.

Analysis of the bill rate:

	£
Bill rate, item 36C	80.61
15% OHP (i.e. deduct 15/115)	10.51
	70.10
Labour placing, say 4.00 man/hrs @ £5.35*	21.40
Cost of concrete at mixer per m³	48.70

Build-up of new rate:

	£
Cost of concrete at mixer per m³	48.70
Labour placing, say 3.25 man/hrs @ £5.35	17.39
	66.09
15% OHP	9.91
Required rate	76.00

---

* All-in labour rate calculated in accordance with the principles of the IOB *Code of Estimating Practice.*

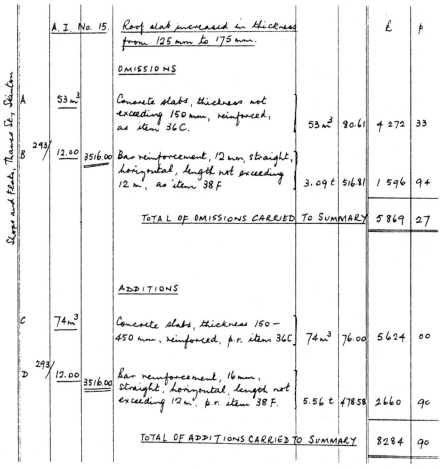

Slopes and Flats, Thames St., Skinton

	A. I. No. 15.		Roof slab increased in thickness from 125 mm to 175 mm.			£	p
			OMISSIONS				
A	53 m³		Concrete slabs, thickness not exceeding 150 mm, reinforced, as item 36C.	53 m³	80.61	4 272	33
B	293/12.00	3516.00	Bar reinforcement, 12 mm, straight, horizontal, length not exceeding 12 m, as item 38F	3.09 t	516.81	1 596	94
			TOTAL OF OMISSIONS CARRIED TO SUMMARY			5 869	27
			ADDITIONS				
C	74 m³		Concrete slabs, thickness 150 – 450 mm, reinforced, p.r. item 36C	74 m³	76.00	5 624	00
D	293/12.00	3516.00	Bar reinforcement, 16 mm, straight, horizontal, length not exceeding 12 m, p.r. item 38F.	5.56 t	478.58	2 660	90
			TOTAL OF ADDITIONS CARRIED TO SUMMARY			8 284	90

Figure 8.2

Reinforcement

An increase in the diameter of reinforcing bars results in a lower cost per tonne of both the material and the labour content.

Analysis of the bill rate:

	£
Bill rate, item 30F, 12 mm bars, per tonne	516.81
15% OHP	67.41
	449.40
Cost of bars	319.34
Labour, bending and fixing	130.06

The balance of cost is reasonable for the required labour, thus indicating that the analysis of the bill rate is reasonable.

Build-up of new rate:

	£
16 mm high tensile steel bars, per tonne	303.72
Labour, bending and fixing, say	112.44
	416.16
15% OHP	62.42
Required rate	478.58

Formwork

No adjustment to the formwork rates is necessary. Although that to the edge of the roof is increased in height, it is still in the same height classification as that to the thinner slab, and the rate therefore remains the same (*See* SMM7, E20).

AI No. 23 Larger windows in living rooms in flats

Only the items affected by the variation were omitted from the original dimensions, i.e. the 'vertical' items. It should be noted that the deduction of walling for the larger windows can be taken-off either as 'deducts' items in the additions, or as omissions items. In the former case, as in the example, the results show negative quantities and so these items have been billed as omission items with positive quantities.

The windows item in the bills was priced at a rate of £97.00. This was analysed and a rate found for the larger windows as follows:

	£
Bill rate, item 64C	97.00
15% OHP	12.65
	84.35
P.c. of window, delivered	75.69
Labour, handling and fixing	8.66

						£	p.
	A.I. No. 23		<u>Larger windows in Living Room</u>.				
			OMISSIONS				
A	10/ 1	10	Softwood window, 1800 x 1200 mm, as D.J. Joinery Ltd's catalogue, ref. WM 1812 N, as item 93C	10 no.	97.00	970	00
B	10/2/3/ 0.50 0.35	10.50	Standard plain glass, panes area 0.15 – 4.00 m², as item 103B				
	10/3/ 0.55 0.37	6.11 16.61		17 m²	25.98	441	66
C	10/ 1.80 1.20	21.60	Two undercoats, one gloss finishing coat on ready-primed wood, glazed windows as item 108A.	22 m²	3.94	86	68
			&				
D			Ditto external as item 109C	22 m²	4.09	89	98
E	10/ 1.80 0.30	5.40	Walls, 102 mm thick, facework one side as item 1C, A.I. No.8	5 m²	30.71	153	55
			&				
F			Ditto, 100 mm thick, lightweight concrete blocks, as item 42G	5 m²	12.04	60	20
			&				
G			Forming cavities in hollow walls as item 43B.	5 m²	0.85	4	25
			carried forward			1 806	32

Shops and Flats, Thames St., Skinton

Figure 8.3

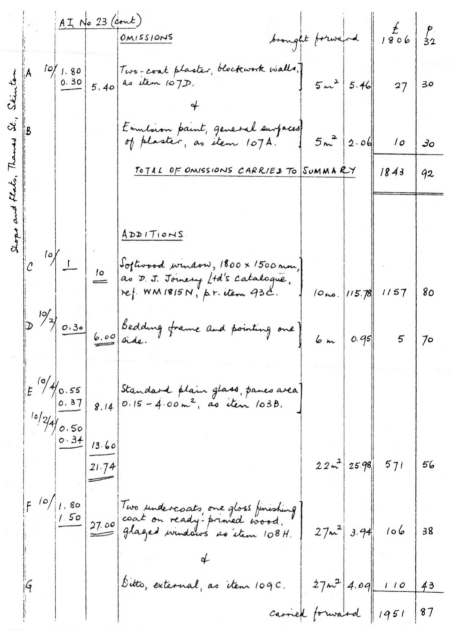

						£	p
			A.I. No 23 (cont)				
			OMISSIONS	brought forward		1806	32
A	10/	1.80	Two-coat plaster, blockwork walls, as item 107D.	5 m²	5.46	27	30
		0.30 5.40					
			⧓				
B			Emulsion paint, general surfaces of plaster, as item 107A.	5 m²	2.06	10	30
			TOTAL OF OMISSIONS CARRIED TO SUMMARY			1843	92
			ADDITIONS.				
C	10/	1 = 10	Softwood window, 1800 × 1500 mm, as D. J. Joinery Ltd's catalogue, ref. WM1815N, pr. item 93C.	10 no.	115.78	1157	80
D	10/2/	0.30 6.00	Bedding frame and pointing one side.	6 m	0.95	5	70
E	10/4/	0.55	Standard plain glass, panes area 0.15 – 4.00 m², as item 103B.				
		0.37 8.14					
	10/2/4/	0.50					
		0.34 13.60					
		21.74		22 m²	25.98	571	56
F	10/	1.80	Two undercoats, one gloss finishing coat on ready-primed wood, glazed windows as item 108H.	27 m²	3.94	106	38
		1.50 27.00					
			⧓				
G			Ditto, external, as item 109C.	27 m²	4.09	110	43
			carried forward			1951	87

(left margin, vertical): Shops and Flats, Thames St., Skinton

Figure 8.3 continued

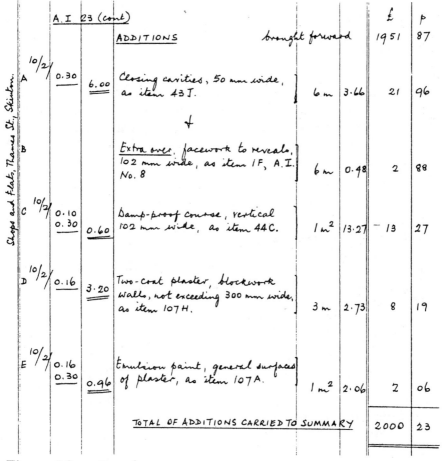

							£	p
			A.I 23 (cont)					
			ADDITIONS	brought forward			1951	87
Shops and flats, Thames St., Skinton.	A 10/2/	0.30 <u>6.00</u>	Closing cavities, 50 mm wide, as item 43 J.		6 m	3.66	21	96
			✝					
	B		Extra over, facework to reveals, 102 mm wide, as item 1F, A.I. No. 8		6 m	0.48	2	88
	C 10/2/	0.10 0.30 0.60	Damp-proof course, vertical 102 mm wide, as item 44 C.		1 m²	13.27	— 13	27
	D 10/2/	0.16 <u>3.20</u>	Two-coat plaster, blockwork walls, not exceeding 300 mm wide, as item 107 H.		3 m	2.73	8	19
	E 10/2/	0.16 0.30 0.96	Emulsion paint, general surfaces of plaster, as item 107 A.		1 m²	2.06	2	06
			TOTAL OF ADDITIONS CARRIED TO SUMMARY				2000	23

Figure 8.3 continued

The labour for the larger window was increased by 20% and added to the p.c. of the larger window, as follows:

	£
P.c. window, delivered	90.29
Labour, handling and fixing	10.39
	100.68
15% OHP	15.10
Required rate	115.78

The remaining rates required to price the additions items are either bill rates or those used in pricing items in AI No. 8. The latter are the brick facework items for which rates are now contained in the additions items of AI No. 8. The original bill rates for those items are, of course, no longer relevant.

AI No. 28 Glazing to balcony screens

In order to determine a rate for the obscure glazing, the bill rate was analysed and a new rate was then synthesized from the resulting data, thus:

	£
Bill rate, item 103A, per m²	27.85
15% OHP	3.63
	24.22
4 mm float glass, including waste	21.38
Labour	2.84

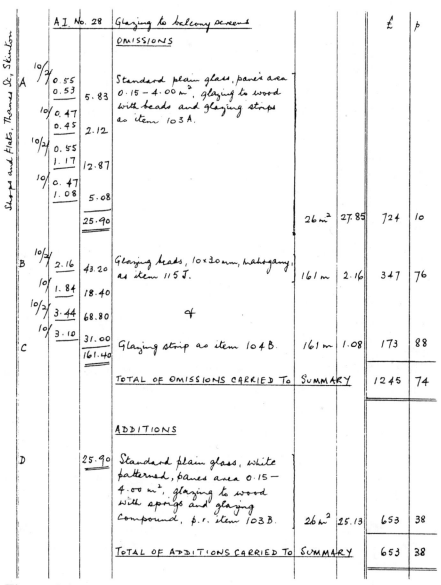

Shops and Flats, Thames St., Skinton

AI. No. 28 | Glazing to balcony screens

OMISSIONS

| | | | | £ | p |

A
10/2/ 0.55
0.53 | 5.83
Standard plain glass, panes area 0.15 – 4.00 m², glazing to wood with beads and glazing strips as item 103 A.

10/ 0.47
0.45 | 2.12

10/2/ 0.55
1.17 | 12.87

10/ 0.47
1.08 | 5.08

25.90 | 26 m² | 27.85 | 724 | 10

B
10/2/ 2.16 | 43.20
Glazing beads, 10 × 20 mm, mahogany, as item 115 J. | 161 m | 2.16 | 347 | 76

10/ 1.84 | 18.40

10/2/ 3.44 | 68.80

4

10/ 3.10 | 31.00

C | 161.40 | Glazing strip as item 104 B. | 161 m | 1.08 | 173 | 88

| TOTAL OF OMISSIONS CARRIED TO SUMMARY | | | 1245 | 74 |

ADDITIONS

D | 25.90 | Standard plain glass, white patterned, panes area 0.15 – 4.00 m², glazing to wood with sprigs and glazing compound, p.r. item 103 B. | 26 m² | 25.13 | 653 | 38

| TOTAL OF ADDITIONS CARRIED TO SUMMARY | | | 653 | 38 |

Figure 8.4

The balance left for labour is reasonable, representing about 0.50 man/hrs (on the basis of an 'all-in' craftsman's rate of £5.47 per hour). So again, the data from the analysis may be used for building-up a new rate, as follows:

	£
4 mm obscure glass, group 2	18.04
Putty and glazing sprigs, say,	0.40
Labour as before + 20%	3.41
	21.85
15% OHP	3.28
Required rate	25.13

AI No. 41 New door to existing store

The omissions relate to the redecoration of the existing window, as provided for in the bills.

The contractor submitted Daywork Sheet No. 6, dated 27 March 1995, showing all the labour and materials expended. The surveyor decided, however, that the new joinery, ironmongery and painting were items that were properly measurable and priceable and should not therefore be valued as dayworks. The carpenter's, painter's, and part of the labourer's time, as well as the relevant materials, have therefore been omitted from the daywork account.

The labour rates and materials prices used in valuing the daywork are those current at the dates when the work was done, i.e. 24–25 March 1995. The labour rates have been calculated in accordance with Section 3 of the *Definition of Prime Cost of Dayworks carried out under a Building Contract*. The respective percentage additions are those tendered by the contractor in the bills (see Appendix A, p. 338).

An addition item for which there was no directly comparable bill item was the fixing item for the push plate. A rate was used which was equivalent to that used for other surface-fixed items of ironmongery – e.g. numerals, hat-and-coat hooks – priced at £1.15.

Shops and Flats, Thames St., Skinton

	A. I. No. 41	New door to existing Store.			£	p
		OMISSIONS				
A	0.68 0.90	One undercoat, one gloss finishing coat on existing wood glazed window as item 108 E	1 m²	3·72	3	72
	0.57	&				
B		Ditto, external, as item 109 F	1 m²	3·84	3	84
		TOTAL OMISSIONS CARRIED TO SUMMARY			7	56
		ADDITIONS				
C	1	Door, 838 × 1981 × 44 mm, exterior quality ply-faced, p.r. item 96 C	1 No.	41·31	41	31
		&				
D		Door frame, 63 × 88 mm, rebated, for door size 838 × 1981 mm, p.r. item 95 D.	1 No.	66·98	66	98
E	7 3 6	Frame tie as item 43 L	6 No.	0.77	4	62
		838 1981 ×28·19 5638 7/7/7 50 400 6038				
F	6.04	Bedding frame and pointing one side as item 98 D.	6 m	0.95	5	70
G	1½	Pair butts, 100 mm, steel, as item 114 D.	1½ No.	4.30	6	45
		carried forward			125	06

Figure 8.5

Shops and Flats, Thames St., Skinton

						£	p.
		A.I. No. 41 (cont)					
		ADDITIONS					
			brought forward			125	06
A	1	Fixing only mortice deadlock as item 114 M		1 No.	6.98	6	98
		&					
B		Ditto pull handle, 75 x 200 mm, as item 114 H.		1 No.	1.44	1	44
		&					
C		Ditto push plate, 75 x 200 mm.		1 No.	1.15	1	15
		&					
D		Ditto door closer as item 114 Q.		1 No.	5.76	5	76
E	0.88 2.03	1.79	One coat primer, two undercoats, one gloss finishing coat, wood general surfaces, as item 108 D.	2 m^2	4.46	8	92
		&					
F		Ditto, external, as item 108 K.		2 m^2	4.76	9	52
G	4.90	4.90	Ditto, wood general surfaces, not exceeding 300 mm girth, as item 108 G.	5 m	1.82	9	10
		&					
H		Ditto, external as item 109 B.		5 m	2.02	10	10
J		Daywork account as daywork sheet no. 6 dated 27/3/95:					
		Labour 198.40 50% addition 99.20				297	60
		Materials 56.42 20% addition 11.28				67	70
		TOTAL ADDITIONS CARRIED TO SUMMARY				543	33

Figure 8.5 continued

BEECON LTD. River Road, Skinton   Contract: Shops and Flats, Thames St., Skinton.

Work commenced: 24 March 1995
Work completed: 25 March 1995

Daywork Sheet No. 6

Date: 27 March 1995
A I No. 41

Take out window from existing Store. Alter opening as necessary to form door opening and provide and build in Catnic lintel. Provide and fix 838 x 1981 mm door and frame, fit deadlock, handles and closer (provided out of p.c. sum for ironmongery).

Labour	Hrs.	Rate	£
Bricklayer	20	5.35	107.00
Carpenter	8	–	–
Labourer	20 / 24	4.57	91.40
Painter	6	–	–
			198.40

Materials	Qty	Rate	£
Facing bricks	60 no.	22.70	13.62
450 x 225 x 100 mm Thermalite blocks	18 no.	55p.	9.90
Mortar	.05 m3	34.62	1.73
Catnic steel cavity lintel 1500 mm	1 no.	27.92	27.92
102 mm Damp-proof course	5 m	65p.	3.25
838 x 1981 mm External quality ply-faced flush door	1 no.	–	–
Door frame set	1 no.	–	–
Undercoat paint	.25 lit.	–	–
Gloss paint	.25 lit.	–	–
			56.42

Plant

The information shown on this sheet has been verified (except rates and prices).

Site Agent's signature _W. Jones_   Architect's/Architect's representative's signature _A. Brous._

Figure 8.6 Example of daywork sheet

The door item was priced *pro rata* to item 96C, which was for a similar type of door but size 762 × 1981 mm. The rate was arrived at as follows:

	£	£
Bill rate, item 96C		39.12
Additional prime cost of 838 mm door	1.54	
Additional labour allowance	0.36	
	1.90	
15% OHP	0.29	2.19
Required rate		41.31

The door frame item was priced *pro rata* to bill item 95D which was for a door frame size 762 × 1981 mm, taking into account also other items for frames. The items in the bills were

	£
50 × 88 mm Frame for door size 762 × 1981 mm	51.72
50 × 88 mm Frame for door size 838 × 1981 mm	53.15
63 × 88 mm Frame for door size 762 × 1981 mm	65.17

On a size proportion basis, the rate was determined at £66.98. The remaining items were priced at bill rates.

AI No. 48 Omission of bookcase units

The omissions formed a complete section of the original taking-off and cols. 325–336 of the dimensions were marked 'X' at the beginning and end of the section for ease of reference. The items involved formed a separate group in the Furniture/Equipment section of the bills and were omitted in the variation accounts as a group together with the related decoration items.

The consequential additions items were all deduction items in the original dimensions, and, upon identification for omissions purposes, they were converted into additions items with positive quantities (*see* comments under AI No. 23, p. 126).

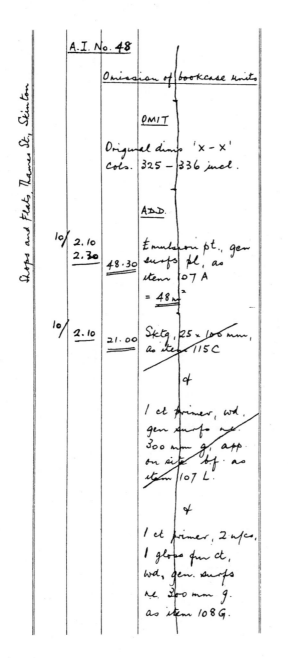

Figure 8.7

						£	p.
	A.I. No. 48						
	Omission of bookcase units						
	OMISSIONS						
A	Items 110 A - 110 K inclusive.					3559	68
B	Item 112 C					241	92
C	Item 112 F					18	36
D	Two coats polyurethane clear finish wood, general surfaces, as item 107 J	22	m²	1.88		41	36
E	Ditto, not exceeding 300 mm girth, as item 107 K	66	m	0.71		46	86
	TOTAL OMISSIONS CARRIED TO SUMMARY					3 908	18
	ADDITIONS						
F	Skirtings, 25 x 100 mm, softwood as item 115C	21	m	2.81		59	01
G	Emulsion paint, general surfaces of plaster as item 107 A	48	m²	2.06		98	88
H	One coat primer, wood general surfaces not exceeding 300 mm girth, applied on site prior to fixing, as item 107 L.	21	m	0.35		7	35
J	One coat primer, two undercoats, one gloss finishing coat, wood general surfaces, not exceeding 300 mm girth as item 108 G.	21	m	1.82		38	22
	TOTAL ADDITIONS CARRIED TO SUMMARY					203	46

Shops and flats, Thames St., Skinton

Figure 8.7 continued

References

1. *Standard Method of Measurement of Building Works, Seventh Edition* (London: RICS Books, 1988) p. 11.
2. *Definition of Prime Cost of Daywork carried out under a Building Contract* (London: RICS Books, 1975).
3. RICS, Questions and Answers, *Chartered Quantity Surveyor*, Vol. 8, No. 8, 1986, p. 8.
4. SIMS, J. Contracts, *Building*, 1975, Vol. 228, No. 6, p. 83; and Vol. 229, No. 49, p. 72.
5. JCT, *Consolidated Main Contract Formula Rules* (London: RIBA Publications Ltd, 1987).
6. BCIS, *Guide to Daywork Rates* (Kingston-upon-Thames: RICS, 1988).
7. KEATING, D., *Law and Practice of Building Contracts* (London: Sweet & Maxwell, 1978), p. 315; WALKER-SMITH, *The Standard Forms of Building Contract* (London: Charles Knight & Co. Ltd, 1971), p. 43.

9

Fluctuations – traditional method of reimbursement

Introduction

Any contract, other than a prime cost contract, will be either 'firm price' or 'fluctuating price'. *Firm price contracts* are those where reimbursement of changes in costs is limited to those relating to statutory contributions, levies, and taxes. *Fluctuating price contracts* are those containing provisions for reimbursement of changes in a wide range of labour costs and materials prices in addition to statutory costs. It should be recognized that the term 'fluctuations' includes both increases and decreases and that they may lead to payments becoming due to the contractor from the Employer and vice versa.

The JCT Form contains three different provisions, any one of which may be incorporated in a contract by deleting the other two clause references in the Appendix to the Conditions of the Contract. These provisions are:

(*a*) adjustment of the contract sum limited to fluctuations in statutory contributions, levies and taxes, as specified in clause 38;
(*b*) adjustment of the contract sum in respect of fluctuations in labour and materials costs and in statutory contributions, levies and taxes, as specified in clause 39;

(c) adjustment of the contract sum in respect of fluctuations as in (b) above by the use of price adjustment formulae, as specified in clause 40.

(a) and (b) above relate to what may be called the *traditional method* of reimbursement of the actual amount of fluctuations allowed under the respective clauses of the contract, and (c) to the *formula method* of calculating by use of formulae a sum to compensate for loss due to fluctuations. This method is discussed in Chapter 10.

Traditional method

The object of the traditional method is to ascertain the actual amount of the fluctuations in costs and prices within the scope of the appropriate clauses of the contract, as given above. That amount will often be less, sometimes much less, than the total amount by which costs and prices have really fluctuated, due to the limitations imposed by the operation of the contract clauses. Those costs and prices may be considered under the three main headings referred to in the JCT Form.

Statutory contributions, levies, and taxes relative to 'workpeople' and others

Clauses 38 and 39.2 are similar and provide for the reimbursement of fluctuations occurring after the Base Date in (i) the amounts of any statutory contribution, levy or tax which, at the Base Date, the contractor is liable to pay or (ii) the amount of any refund of any such contribution, levy or tax which is receivable by the contractor at the Base Date, as an employer of workpeople (as defined in clauses 38 and 39). The Base Date is defined as the date stated in the Appendix to the Conditions.

Examples of these costs are: employers' National Health Insurance contributions and contributions under the Social Security Act 1986. The wording of the clauses provides not only for changes in contributions, etc., operative at the Base Date but also for new types of contribution or tax which may be imposed and for existing ones which may be abolished.

The provisions of clause 38 relate not only to operatives employed on site but also to:

(*a*) persons who are directly employed elsewhere by the contractor in the production of materials or goods for the Works, such as joinery shop workers; and
(*b*) persons, such as timekeepers and site surveyors, who, although employed by the contractor on site, are not 'workpeople' as defined in the clause. Amounts in respect of this class of persons are the same as for a craftsman, but in proportion to the amount of time spent on site per week, within certain specified limitations.

Labour costs in addition to statutory costs

Clause 39.1 deals with fluctuations in labour costs other than those imposed by statute. In this sub-clause, the basis of the prices in the bills or schedule of rates is first defined and then it is provided that any increases or decreases within the prescribed limits shall be paid to or allowed by the contractor.

The basis of adjustment is 'the rules or decisions of the National Joint Council for the Building Industry' and these are embodied in the *National Working Rule Agreement*. This agreement sets out basic wage rates, extra payments to which operatives may be entitled and various matters relating to working conditions. It also includes the provisions of the Annual Holidays Agreement and the amounts of weekly holiday credits for each operative for which an employer is liable. Specific reference is made in clause 39.1.1 and 39.1.2 to employers' liability insurance and third party insurance, which brings the respective premiums within the reimbursement provisions, for the reason that the premiums are usually calculated as a percentage of a contractor's wages bill.

It should be noted that the rates of wages referred to are those 'in accordance with the rules or decisions of the National Joint Council'. Where operatives are paid at higher wage rates than the nationally-negotiated rates, the basis of the prices in the contract bills or schedule of rates is no longer as stated in clause 39.1.1 and therefore clause 39.1.2 cannot follow.[1]

Operatives employed by 'domestic' (non-nominated) sub-contractors (whether on a labour-only basis or not) rank for recovery of fluctuations in the same way as, and on a similar basis to, the main contractor's directly employed labour,[2] provided that the contractor has complied with the requirement of clause 39.4.1.

The classes of persons coming within the scope of clause 39 are the same as those to which clause 38 applies, as described above.

Materials costs

Where clause 39 applies

The basis of adjustment is a list of materials (and goods) with their basic prices. Both materials and prices may be entered by the contractor on a blank form attached to the bills (see Appendix A, p. 341) at the same time as they are priced or the form may be required to be filled in and returned with the Form of Tender. Sometimes, the principal materials may be listed on the form by the surveyor, the contractor adding the prices when tendering. The prices on the list are intended to be market prices current at the Base Date and once the contract is signed, they must be regarded as such anyway. Any fluctuations, to be allowable for recovery, must first, relate to materials named on the list and secondly, be due solely to changes in the market prices. Such changes may arise from, or include, a change in tax or duty on a material or from either the abolition of existing duty or tax or the imposition of new types of duty or tax.

Sometimes, claims are made in respect of changes in price which are due to other causes than changes in market prices. These may be: buying in part loads or small quantities, obtaining materials from another supplier at less favourable discounts, or because the material is not the same as, although similar to, that on the Basic Prices List. Adjustments to the contract sum should not be made for fluctuations in prices due to these or similar causes.

Clause 39.7.2 is specific that timber used in formwork ranks for fluctuations recovery (provided timber is on the Basic Prices List) although it is not part of the permanent work. Other consumable stores, plant and machinery do not, with the exception of electricity and, where specifically stated on the Basic Prices List, fuels.

The inclusion in a claim of materials drawn from the contractor's stores sometimes presents a problem, if the contractor is unable to produce invoices for them. Clause 39.3.2 is specific that the sum to be paid to or allowed by the contractor is to be the difference between the prices on the Basic Prices List and the market prices current when the materials were bought, and which were payable by the contractor. Some surveyors will not include materials from stock for which the contractor cannot produce an invoice. Others will include them on assumed purchase prices equivalent to those being paid for other quantities of the same materials.

Materials used by 'domestic' (non-nominated) sub-contractors rank equally with those used directly by the general contractor and on the same basis, provided that the contractor has complied with the requirement of clause 39.4.1.

The surveyor should check the total quantities of each material listed in the claim to see that they are not in excess of those reasonably required, remembering to make allowances for reasonable losses due to wastage and theft. Any excess quantities should, of course, be excluded.

Where clause 38 applies

The basis of adjustment is a list (but no prices) attached to the contract bills, of materials, goods, electricity and, where specifically stated in the contract documents, fuels. Any adjustments allowable are limited to changes in types or rates of duty or tax relative to any of the materials or goods, etc., listed, and to price changes due to the abolition of a type of duty or tax or to the imposition of a new one. The comments made in the preceding section apply here also.

Nominated sub-contractors and nominated suppliers

Claims for fluctuations from nominated sub-contractors and nominated suppliers are specifically excluded from the provisions of clauses 38 and 39, being dealt with under the provisions of the respective sub-contracts and contracts of sale. If the sub-contracts and contracts of sale contain clauses which provide for the recovery of fluctuations, then the amounts properly recoverable under them will be payable to or allowable by the general contractor, as the case may be.

With regard to nominated sub-contracts, where the Standard Form of Nominated Sub-contract (NSC/C) is used, clauses 4A, 4B, and 4C inclusive are parallel to clauses 38 to 40 of the main contract. Any proper claims for reimbursement of fluctuations under the appropriate sub-contract clauses will then form part of the amounts payable to the general contractor by the Employer under clause 30.6.2.6 of the main contract. The surveyor will need to know, therefore, in order to check a sub-contractor's account, what the details are in Part 2 of the Standard Form of Nominated Sub-Contract Tender (NSC/T) attached to the sub-contract.

With regard to contracts of sale between the contractor and nominated suppliers, the provisions of each contract will determine what fluctuations (if any) are properly recoverable. Those that are, as in the case of sub-contracts, will become part of the amounts 'properly chargeable to the Employer' under clause 30.6.2.8.

Additions to net sums recoverable

The sums calculated for fluctuations are not subject to any profit addition (*see* clauses 38.4.6 and 39.5.6 of the JCT Form) but an addition should be made of the percentage stated in the Appendix to the Conditions as provided in clauses 38.7 and 39.8. This addition applies to all fluctuation amounts recoverable under the provisions of clauses 38 and 39.

The percentage addition was introduced in order to go some way towards meeting the objections of contractors that only a small proportion of the true amounts of the increased costs actually incurred was being recovered.

Exclusions

The following kinds of work are excluded from the fluctuations provisions of clauses 38 and 39 of the JCT Form.

(a) Work paid for on a daywork basis. This is because the rates used for valuing daywork should be rates current at the time the work is done. Those rates are used (and not tender date rates) because of the percentage additions made to the prime costs (see p. 116).

(b) Work carried out by a nominated sub-contractor and materials and goods supplied by a nominated supplier. As stated above, fluctuations relative to work or materials in this category are dealt with under clauses 35 and 36.

(c) Work carried out by the contractor which is the subject of a prime cost sum and for which the contractor submitted a tender under clause 35.2. The reason here is the same as in (b) above.

(d) Changes in value added tax charged by the contractor to the Employer.

Recording and checking fluctuations claims

It is in the interests of the contractor and the surveyor that a satisfactory system of recording and submitting the details of fluctuations claims be agreed at the start of the contract. This means that the system employed should facilitate the checking of the details from original invoices, wages sheets, etc. The amount of checking required can be considerable and the more systematic the contractor is in recording the data and preparing his claim, the sooner the checking and computation of the sums payable can be completed.

In order that amounts due may be paid without delay, the contractor should present claims at regular intervals, say, weekly or fortnightly. All supporting documents should be provided for the surveyor's inspection when checking the claim and it is useful to have the signature of the clerk of works, if there is one, on time sheets to certify the correctness of the names and hours shown. Alternatively, if a standard claim form is used to record the data taken from time sheets, provision may be made on the form for certification of the correctness of that data.

Figure 9.1 shows a summary of part of the materials claim relating to the contract of which particulars are given in Appendix A. The claim was prepared in accordance with the particulars shown in the Appendix to the Conditions of the contract, i.e. under the provisions of clause 39. Only the increases in the cost of cement are shown, being part only of the complete materials fluctuations claim. The higher price of the May delivery was due to the small quantity and not to an increase in market price. It was therefore reduced by the surveyor to the rate appropriate to full loads.

BEECON LTD, River Road, Skinton   Contract Shops and Flats, Thames St, Skinton

MATERIALS FLUCTUATIONS CLAIM

Material Cement
Basic price *£77.00*

Sheet No. 1

Supplier	Invoice No.	Date	Quantity	Invoice price	Decrease	Increase	Total decrease	Total increase
Grey Square Cement Co. Ltd.	DF897	16.11.94	20 tonnes	79.25	—	2.25	—	45.00
" " "	LX1376	11.1.95	10 "	79.25	—	2.25	—	22.50
Greenthorne Cement Co. Ltd.	G2736X	15.3.95	12 "	79.25	—	2.25	—	27.00
" " "	K1556Z	5.4.95	10 "	79.25	—	2.25	—	22.50
" " "	K2485W	2.5.95	2 "	79.25 ~~87.55~~	—	2.25 ~~10.55~~	—	4.50 ~~21.10~~
Grey Square Cement Co. Ltd.	PR884	18.7.95	6 "	79.25	—	2.25	—	13.50
" " "	DL1285	20.9.95	6 "	82.40	—	5.40	—	32.40
Net total carried to summary							—	167.40 ~~184.00~~

Figure 9.1 Example of materials fluctuation claim

Payment

Payment of amounts recoverable is required to be made in the next interim certificate after ascertaining, from time to time, the amounts due. Such amounts are not subject to retention. The total amount of fluctuations will also be included in the final account as an addition or deduction, as the case may be.

Fluctuations after expiry of the contract period

A frequent cause for dispute in the past was whether the contractor was entitled to be paid for increases in costs relative to the period between the 'Date for Completion' (i.e. the date, as stated in the Appendix or extended, by which the Contract should have been completed) and the date of actual completion. It is now clearly established that fluctuations amounts are recoverable until the date of actual completion, that is, including the period after the date by which the contract should have been completed. The decision of the Court of Appeal in the case of *Peak Construction (Liverpool) Limited v McKinney Foundations Limited (1970), 69 LGR 1*, settled the matter and although the contract in that case was not based on the JCT Form, the Joint Contracts Tribunal has accepted the judgment as applying to the JCT Form.[3]

The JCT Form, in clauses 38.4.7 and 39.5.7, 'freezes' the amount of fluctuations, however, at the levels which are operative at the Completion Date (as stated in the contract or as extended). Consequently, although fluctuations should be calculated up to the date of actual completion, such calculations should be made on the basis of the changes in costs and prices occurring up to, but not after, the Completion Date.

Other non-recoverable fluctuations

It has already been pointed out (*see* p. 142) that where a contractor pays operatives at rates above the nationally-negotiated rates, the provisions of clause 39 no longer apply. That clause applies only to 'rates of wages and other emoluments and expenses . . . payable . . . in accordance with the rules or decisions of the National Joint Council for the Building Industry . . .'. Any fluctuations in such inflated rates of wages are not recoverable, even where the fluctuations are related to nationally-agreed rates.

The non-productive hours of overtime (even when expressly authorized) should not be included in allowable fluctuations payments. Only the actual hours worked rank for reimbursement.

Also excluded is the time spent on making good defects under the provisions of clause 17. Such work is required to be done entirely at the contractor's expense.

References

1. The Query Sheet, *Chartered Surveyor, BQS Quarterly*, Vol. 2, No. 4, Summer 1975, p. 56.
2. Questions and Answers, *Chartered Surveyor, BQS Quarterly*, Vol. 4, No. 1, Autumn 1976, p. 15.
3. The Query Sheet, *Chartered Surveyor, BQS Quarterly*, Vol. 2, No. 2, Winter 1974, p. 27.

10

Fluctuations – formula method of reimbursement

In contrast to the traditional method of reimbursement of fluctuations, the object of the formula method is to calculate a sum or sums which will compensate the parties to the contract for loss incurred due to increases and decreases in costs. No attempt is made to calculate the *actual* amount of loss involved. Consequently, the sums recoverable by the formula method will differ from those recoverable under the traditional method and will usually be greater.[1]

The formula method is incorporated in the 'with quantities', 'without quantities' and 'with approximate quantities' variants of the JCT Form, in both Local Authorities and Private Editions, by means of clause 40 as the alternative to clauses 38 and 39. It may also be used for schedule of rates contracts.

The JCT Form refers in clause 40.1 to a separate document called the *Formula Rules*[2] which sets out the formulae and defines their use. By virtue of the provisions of clause 40, the *Formula Rules* are made part of the Standard Form and, therefore, contractually binding on the parties.[3]

The application of the formulae is by means of indices, published monthly by the Department of the Environment[4], which measure the fluctuations in costs of building work classified in sixty categories.* Figure 10.1 gives the list of Series 3 work categories. The indices are intended to be applied monthly

---

* Work Groups may be used instead. A Work Group is 'a work category or combination of work categories as defined in the contract bills'. Wherever the term 'work category' is used hereafter, the words 'or work group' are to be implied where appropriate.

Series 3 Building Formula
Work Categories

01 Demolitions
02 Excavation & Disposal
03 Filling: Imported, Hardcore & Granular
04 Piling: Concrete
05 Piling: Steel
06 Concrete: In Situ
07 Concrete: Formwork
08 Concrete: Reinforcement
09 Concrete: Precast
10 Brickwork & Blockwork
11 Stonework
12 Softwood Carcassing & Structural Members
13 Metal: Decking
14 Metal: Miscellaneous
15 Cladding & Covering: Glazed
16 Cladding & Covering: Fibre Cement
17 Cladding & Covering: Coated Steel
18 Cladding & Covering: Plastics
19 Cladding & Covering: Glass Reinforced Plastics
20 Cladding & Covering: Slate & Tile
21 Cladding & Covering: Lead
22 Cladding & Covering: Aluminium
23 Cladding & Covering: Copper
24 Cladding & Covering: Zinc
25 Waterproofing: Asphalt
26 Waterproofing: Built Up Felt Roofing
27 Waterproofing: Liquid Applied Coatings
28 Boards, Fittings & Trims: Softwood
29 Boards, Fittings & Trims: Hardwood
30 Boards, Fittings & Trims: Manufactured
31 Linings & Partitions: Plasterboard
32 Linings & Partitions: Self Finished
33 Suspended Ceilings
34 Raised Access Floors
35 Windows & Doors: Softwood
36 Windows & Doors: Hardwood
37 Windows & Doors: Steel
38 Windows & Doors: Aluminium
39 Windows & Doors: Plastics
40 Ironmongery
41 Glazing
42 Finishes: Screeds
43 Finishes: Bitumen, Resin & Rubber Latex Flooring
44 Finishes: Plaster
45 Finishes: Rigid Tiles & Terrazzo Work
46 Finishes: Flexible Tiles & Sheet Coverings
47 Finishes: Carpets
48 Finishes: Painting & Decorating
49 Sanitary Appliances
50 Insulation
51 Pavings: Coated Macadam & Asphalt
52 Pavings: Slab & Block
53 Site Planting
54 Fencing
55 Pipes & Accessories: Spun & Cast Iron
56 Pipes & Accessories: Steel
57 Pipes & Accessories: Plastics
58 Pipes & Accessories: Copper
59 Pipes & Accessories: Aluminium
60 Pipes & Accessories: Clay & Concrete

Figure 10.1 Price adjustment work categories

to the sums included in interim valuations, thus ensuring that the payment of fluctuations is made promptly. The index number for each work category is based upon price movements of a number of selected items of labour, materials and plant which are calculated from a range of contracts.

It follows, therefore, that the sums produced by the application of the indices to the prices in interim valuations must necessarily reflect a *general* fluctuation in prices and cannot relate to *actual* fluctuations in labour, materials and plant costs applicable to a particular contract.

Preparatory work necessary before using the formulae

The *Formula Rules* give a formula for application to each work category (Rule 9), a second one relative to *balance of adjustable work* (i.e. work ranking for adjustment but which is not capable of allocation to any work category) (Rule 26), and a third formula for use for valuations after Practical Completion and in the Final Certificate (Rule 28). After giving corresponding formulae for use with work groups (Rules 29 and 38), five formulae applicable to five types of specialist engineering installations are given (Rules 50, 54, 58, 63 and 69).

In order to apply the appropriate formula to each work category, all the measured work items in the bills of quantities must be allocated to work categories. This should be done before tenders are invited, either by annotating the bills or by preparing schedules. Then, when interim valuations are done, the value of work included in them must be allocated accordingly. Annotating bills may be done by giving the work category number against each item in a column next to the item reference column.

There will be some work or goods, the value of which is included in a valuation, which cannot easily (if at all) be allocated to work categories or which is excluded from adjustment and this is dealt with as discussed below.

Balance of adjustable work

This is the balance of main contractor's work which ranks for adjustment but which is not allocated to work categories or to which an engineering formula does not apply. Examples are preliminaries, water for the works, insurances, attendances on nominated sub-contractors. The amount of the adjustment to the balance of adjustable work for any valuation period is in the same proportion as the total amount of adjustment of work allocated to work categories bears to the total value of work in all work categories.

Multiple category items

Sometimes a bill item, such as an alterations item in a 'Works on Site' bill, may relate to work falling within more than one work category. Such an item is allocated to the work category to which the largest part of the item relates.

Provisional sums

Provisional sums for work which will be paid at current prices (e.g. dayworks, work carried out by statutory undertakers and other public bodies), are not allocated to work categories but are separately identified as not subject to adjustment.

Provisional sums for work to be carried out by the general contractor are not allocated to work categories in the bills, but when the work has been executed, the measured items set against the provisional sum will then be allocated to appropriate categories or, if relating to specialist work, will be adjusted using the appropriate formula.

Fixing of items supplied either by nominated suppliers or by the Employer

'Fix only' items may be dealt with either by (a) allocating to an appropriate work category; (b) including in the balance of adjustable work; or (c) applying an index specially created for the purpose. Whichever method is used must have been previously specified in the bills or agreed with the contractor in writing.

Variations

All additions items priced at bill rates or analogous rates or at rates current at the Base Date will rank for adjustment and should be allocated to work categories. Variations valued on a daywork basis or otherwise valued at rates or prices current at the date of execution should be excluded from the operation of the formulae.

Payments to nominated sub-contractors and nominated suppliers

Such payments, including profit, are excluded from the adjustment. Amounts for attendances on nominated sub-contractors are included in the balance of adjustable work (*see* 'Balance of adjustable work' above).

Unfixed materials and goods

The value of unfixed materials and goods is excluded from adjustment. Only when they have been incorporated into the work will they become adjustable in one of the ways described above.

Claims for loss and/or expense

Claims for extra payments under clauses 26 and 34.3 are similarly excluded as they will be valued at their full value.

Calculation of adjustment amounts

The appropriate formula is applied to each part of the value of work included in a valuation (after deducting the value of any work not subject to adjustment) when it has been apportioned into (*a*) work categories, (*b*) balance of adjustable work, and (*c*) specialist engineering installations. Such application is subject to the following considerations.

Provisional indices

Indices for any month when first published are provisional. Normally, firm indices are published three months later. The calculation of the adjustment amounts initially will also be provisional and will be subject to revision when the next valuation is done following publication of the firm indices.

Base month

The base month from which adjustment is calculated is the calendar month stated in the Appendix to the Conditions, which normally is the one prior to that in which the tender is due for return. The published index numbers for the base month will be used in the application of the formulae.

Valuation period

The value of work carried out during the valuation period is substituted in the formula and this period is defined as commencing on the day after that on which the immediately preceding valuation was done and finishing on the date of the succeeding valuation.

The index numbers for the valuation period are those for the month in which the mid-point of the valuation period occurs. If the valuation period has an odd number of days, then it will be the middle day of the period; if it contains an even number of days, the mid-point will be the middle day of the period remaining after deducting the last day.

Non-adjustable element

The non-adjustable element deduction applies only to contracts based on the local authorities edition of the JCT 'with quantities' Form.

Retention

The fluctuations amounts calculated by the formula method, unlike those ascertained by the traditional method, are subject to retention deductions. The reason for this is that what the formula method really does is to effect an updating of the rates and prices in the contract bills. The original rates and prices when included in valuations are subject to retention and it is logical, therefore, that the updated ones should be also.

Application of the formula rules to work executed after the Completion Date (or extended Completion Date)

Under the provisions of clause 40.7 of the JCT Form the formula is applied to the value of work executed after the Completion Date (or extended completion date) but using the index numbers applicable to the month in which that date falls. This has the same effect as if all the work remaining to be done after the Completion Date had been carried out during the month in which the Completion Date occurred. It should be noted that this provision will not apply if either clause 25 has been amended or the architect has failed to fix or confirm the Completion Date in accordance with the provisions of clause 25.

Adjustment of value of work included in interim valuations issued after the issue of the Certificate of Practical Completion

Rule 28 of the *Formula Rules* gives the formula to be used for adjusting the value of executed work which is included in interim certificates issued after the Certificate of Practical Completion has been issued, such value, in consequence, not being attributable to any valuation period. The formula effects an adjustment to such value in the same proportion as the total fluctuations included in previous certificates bears to the total value of work included in such certificates. Put another way, the adjustment is calculated on the basis of the average fluctuations over the contract period.

This provision would apply to the difference between the actual value and an approximate valuation of variations and/or remeasured work and to the difference between the balance of adjustable work contained in the final account (and therefore in the Final Certificate) and that in respect of which fluctuations have already been included in interim certificates.

Errors

'Mechanical' errors

Possible errors, which may arise in the course of using the formula method, are listed in Rule 5a of the *Formula Rules*. They are:

(*a*) arithmetical errors in the calculation of the adjustment;
(*b*) incorrect allocation of value to work categories or to work groups;
(*c*) incorrect allocation of work as Contractor's Specialist Work;
(*d*) use of an incorrect index number or index numbers.

If any such error has occurred, then the surveyor is required to correct it. Rule 5c requires that, when making the correction, the same index numbers are used as were used when the original calculation was done, except that firm index numbers, if available, should be substituted for provisional ones.

Rule 5b provides for the contractor to be granted access, if he so requests, to the 'working documents' which show the calculations of the fluctuations included in an interim certificate for payment, so that he may satisfy himself that errors of the kind referred to above have not been made.

Errors in preparing a valuation

The *Formula Rules* make no provision for the retrospective correction of errors made in the course of preparing an interim valuation. Such 'errors' include the valuation of provisional items of work at the quantities given in the bills which upon remeasurement are found to be different.

The *New Users Guide* advises that valuations should be as firm and accurate as possible as these will not normally be reassessed at a later date.[5] The Quantity Surveyors (Practice and Management) Committee of the RICS has stated the view that, if errors have occurred in the preparation of a valuation, they should be corrected but that, once firm index numbers have been substituted for provisional ones, the facility for such correction will no longer exist.[6]

When errors of this kind are corrected in the manner envisaged by the RICS Quantity Surveyors Committee, the correction will be effected in the same way as described for 'mechanical' errors, that is, as set out in Rule 5c.

Any further corrections which may remain to be made after firm index numbers have been applied, however, may be dealt with in the same way as the value of work included in interim certificates issued after Practical Completion (*see* above). In other words, adjustment of the errors may be made on the basis of the average of fluctuations over the whole of the contract period.

Examples

The following examples relate to the contract of which particulars are given in Appendix A. The application of the Formula Method, using work categories, is shown as it would have applied had the fluctuations provision been based on clause 40. The Base Month would then have been stated on the Appendix to the Conditions as January 1994. For the purposes of the example, Valuation No. 12 in the examples given at the end of Chapter 12 has been used. The indices used are the 1990 Series (Series 3).

It was necessary, prior to Valuation No. 1, to make preparatory calculations as follows:

(*a*) to ascertain from the bills the value of work in each of the appropriate work categories and this was done on Form A (exclusions are: preliminaries, prime cost and provisional sums, including general contractor's profit and attendances on nominated sub-contractors, builder's work in connection with plumbing and builder's work in connection with electrical installation);

(b) to calculate the balance of adjustable work, as shown on Form B;

(c) to enter on Form WC1/WC2 the published Base Month index numbers for all the work categories involved.

The foregoing information was then available for use when each valuation was done.

When Valuation No. 12 was carried out on 25 March 1995, the following procedure was followed:

(a) the value of the work done *since the last valuation* (including the variation contained in AI No. 62), which could be allocated to work categories, was entered in column 3 on Form WC1/WC2 (*see* p. 163). In column 4, under the heading 'Provisional', the index numbers for March for the relevant categories were entered. In column 5 were entered the adjusted valuation amounts using the expression

$$V \times \frac{I_v}{I_o} \quad \text{where}$$

V = value of work executed in the work category during the valuation period;

I_v = index number for the work category for the month during which the mid-point of the valuation period occurred;

I_o = the work category index number for the Base Month.

(b) the total of items assigned to work categories included in the valuation before adjustment and the amount of the adjustment were then entered at Q and R on Form D (see p. 164). The total value to date of the balance of adjustable work was calculated from the valuation and entered on Form D and from this the balance of adjustable work for Valuation No. 12 was calculated by deducting the amount included in Valuation No. 11. The gross adjustment on the balance of adjustable work was then calculated, as indicated on Form D, and the amount entered in the space beneath the amount in respect of the work categories total. The sum of these two amounts (£1846) was the total of fluctuations for March using the provisional index numbers.

Form A

Allocation of bill of quantities items to work categories

Contract No. ...... 265/6 .... Employer *Skinton Development Co.*

Project ..... *Shops and Flats;* ..... Contractor *Beacon Ltd.*
...... *Thames Street, Skinton:* ....

		Bill of quantities Nos.	2 - 4
Cat No 3	Bill of quantities item reference		Amount £
1	16A – 16F		5685
2	20A – 23G; 138A – 138F; 142A – 142B		26352
3	24B – 26C; 138G – 139J; 142C – 142F		8700
4			
5			
6	27A – 28J		25658
7	29A – 30G		12927
8	31A – 31H		13614
9	32A – 32E		3809
10	34A – 43C; 130A – 136E		56337
11			20909
12	45A – 68D		
13			
14			
15			
16			
17			
18			
19			
20	70A – 77C		9503
21	78A – 78G		3558
22			
23			
24			
25			
26	90D – 91F		15548
27			
28	113A – 114C; 112A – 112D		7284
29			
30			
31			
32			

Total to overleaf £ 209884

Figure 10.2 Fluctuations calculations by formula method

Form A (cont.)

		Total from overleaf £	209884
33			
34			
35	93A – 102F		25604
36			
37			
38			
39			
40	122A – 123L		3725
41	103A – 104C		11625
42	107A – 107D		3122
43			
44	105A – 106P		5807
45			
46	108A – 108f		1756
47			
48	109A – 109S		4836
49	110A – 111K		6174
50	113A – 113L		2431
51	142G – 142M		3111
52	142N – 143G		1477
53			
54	141A – 141F		17726
55			
56			
57	117A – 123C		12941
58	125A – 128E		9127
59			
60	140G – 140K		16451
		Total £	335797
Provisional sums to which formula will apply			6458
		Total L £	342255
Provisional sums to which formula will not apply (including dayworks)		Total M £	14210
PC sums (including main contractor's profit)		Total N £	95256

Figure 10.2 continued

(Part 1)   Summary of Form A

Contract No. _____ 265/6 _____   Employer _____ Skinton Development Co _____

Project _____ Shops and Flats, _____   Contractor _____ Beacon Ltd. _____
_____ Thames Street, Skinton. _____

Bill of quantities No.	Work allocated to categories and provisional sums to which formula will apply (Total L Form A)	Provisional sums to which formula will not apply (including daywork) (Total M Form A)	PC sums including main contractor's profit (Total N Form A)
1			
2			
3	342255	14210	95256
4			
5			
6			
7			
Totals £	X 342255	Y 14210	Z 95256

(Part 2)   Calculation of balance of adjustable work

	Contract sum (before deduction of credit for old materials)	£	495 000
Deduct	(i) Provisional sums to which formula will not apply (including dayworks) – Total Y above	£ 14210	
	(ii) PC sums (including main contractor's profit) – Total Z above	£ 95256	109 466
	Total of contract sum properly subject to price adjustment	£	385 534
Deduct	Value of work allocated to work categories, and provisional sums to which the formula will apply – Total X above	£	342 255
	Balance of adjustable work	£	43 279

Figure 10.2 continued

Form WC1/WC2

Work Category Indices (Base Month) and Calculation of Gross Valuation

Base Month		
JANUARY 1994		

Valuation No	12	Date of Valuation: 25 March 1995
		for month of: March
Contract No.	265/6	
Project: Shops and Flats, Thames Street, Skinton.		

Work Category		Valuation this month (Q) £	PROVISIONAL		FIRM		Work category No 3/
No 3/	Base Month Index		Index for valuation period	Gross valuation including VOP $(3 \times \frac{4}{7})(VT)$ £	Index for valuation period	Gross valuation including VOP $(5 \times \frac{4}{7})(VT)$ £	
1	2	3	4	5	6	7	8
1	·	–		–			1
2	120	·837	123	858			2
3	111	660	116	690			3
6	102	1994	108	2111			6
7	118	1225	126	1308			7
8	105	1015	110	1063			8
9	110	–	109	–			9
10	113	3681	118	3844			10
12	114	–	125	–			12
20	116	–	122	–			20
21	104	–	111	–			21
26	125	2732	133	2907			26
28	113	486	121	520			28
35	115	5745	123	6145			35
40	114	235	120	247			40
41	112	–	115	–			41
42	116	515	119	528			42
44	121	805	125	832			44
50	101	–	110	–			50
51	111	–	123	–			51
54	116	–	121	–			54
57	115	2334	127	2578			57
58	114	2000	130	2281			58
60	119	863	125	907			60
		25127 (0)	(VT)	26819	(VT)		Totals

Figure 10.2 continued

Form D

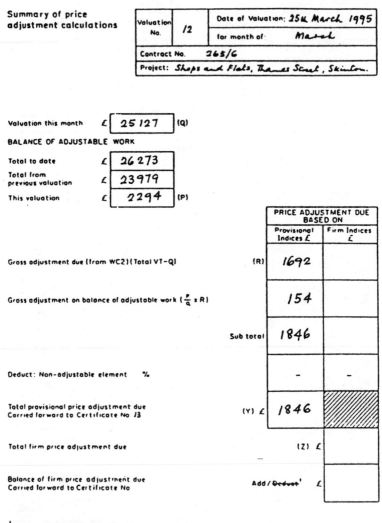

Figure 10.2 continued

Form WC1/WC2

Work Category Indices (Base Month) and Calculation of Gross Valuation

Base Month:		
JANUARY 1994		

Valuation No.	9
Date of Valuation: 23rd December 1994	
for month of: December	

Contract No. 265/6

Project: Shops and Flats, Thames Street, Skinton.

			PROVISIONAL		FIRM		
Work Category		Valuation this month (Q)	Index for valuation period	Gross valuation including VOP $(3 \times \frac{4}{1})(VT)$	Index for valuation period	Gross valuation including VOP $(5 \times \frac{4}{4})(VT)$	Work category No. 3/
No.3/	Base Month Index						
1	2	£ 3	4	£ 5	6	£ 7	8
1							1
2							2
3							3
6	102	1877	108	1987	108	1987	6
7	118	1469	125	1556	125	1556	7
8	105	1349	110	1413	109	1400	8
9	110	770	107	749	108	756	9
10	113	2458	117	3063	117	3063	10
12	114	3047	124	3314	124	3314	12
20	116	2169	121	2262	121	2262	20
28	113	756	121	810	120	803	28
31	124	790	129	822	129	822	31
35	115	3651	120	3810	123	3905	35
41	112	2309	115	2371	116	2391	41
44	121	418	125	432	125	432	44
49	114	5399	115	5446	115	5446	49
50	101	1911	110	2081	110	2081	50
54	116	964	120	997	120	997	54
58	114	1920	122	2055	125	2105	58
60	119	1538	125	1616	125	1616	60
		33295 (Q)	(VT)	34784	(VT)	34936	Totals

Figure 10.2 continued

Form D

Summary of price
adjustment calculations

Valuation No.	9	Date of Valuation 23rd December 1994
		for month of December
Contract No	265/C	
Project: Shops and Flats, Thames Street, Skinton.		

Valuation this month   £ **33 295**   (Q)

BALANCE OF ADJUSTABLE WORK

Total to date   £ **20 274**

Total from   £ **18 627**
previous valuation

This valuation   £ **1 647**   (P)

	PRICE ADJUSTMENT DUE BASED ON	
	Provisional Indices £	Firm Indices £
Gross adjustment due (from WC2) (Total VT-Q) (R)	**1 489**	**1 641**
Gross adjustment on balance of adjustable work $\left(\dfrac{P}{Q} \times R\right)$	**74**	**81**
Sub total	**1 563**	**1 722**
Deduct: Non-adjustable element %	–	–
Total provisional price adjustment due Carried forward to Certificate No. 10 (Y) £	**1 563**	////////
Total firm price adjustment due (Z) £		**1 722**
Balance of firm price adjustment due Carried forward to Certificate No. 13 Add / ~~Deduct~~ * £		**159**

*Delete whichever is not applicable
If Y is >Z then DEDUCT
If Y is <Z then ADD

Figure 10.2 continued

(c) at the same time as provisional index numbers for March became available, firm index numbers were published for December and so the fluctuations total included in Valuation No. 9, carried out on 23 December 1994, calculated on provisional index numbers, could then be adjusted. Accordingly, Form WC1/WC2, used to record the provisional fluctuations calculation for each work category, was then filled in under the column heading 'Firm' (see p. 165). The firm index numbers for December were entered in the 'Index for Valuation Period' column and, in those cases where they differed from the provisional numbers, the revised amount of 'gross valuation including VOP' was calculated. Where there was no difference, the figures were simply carried over from column 5 to column 7. The revised amounts, as appropriate, were then entered on Form D (see p. 166) alongside the provisional ones and the difference in the totals at 'Y' and 'Z' (£159.00) was entered in the last space on the form as an addition for inclusion in Certificate No. 12.

It will be obvious from the foregoing, that it is essential that the record of calculations on Form WC1/WC2 and Form D should be carefully filed away after each valuation, along with Forms A and B.

The calculations described above are shown in Figure 10.2.

References

1. GOODACRE, P. E., *Formula Method of Price Adjustment for Building Contracts* (Reading: College of Estate Management, 1978), p. 81.
2. JCT, *Consolidated Main Contract Formula Rules* (London: RIBA Publications Ltd, 1987).
3. JCT, *Practice Note 17, Fluctuations* (London: RIBA Publications Ltd 1982), para. 8.
4. *Price Adjustment Formulae for Building Contracts, Monthly Bulletin of Indices* (London: HMSO).
5. *Price Adjustment Formulae for Construction Contracts: New Users Guide, 1990 Series of Indices* (London: HMSO, 1995).
6. *Price Adjustment Formulae for Building Contracts* (London: RICS Books 1975), p. 8.

11

Claims

Contractor's claims may be of three kinds:

1 Common law.
2 *Ex gratia.*
3 Contractual.

Common law claims

These arise from causes which are outside the express terms of a contract. They relate to breaches by the Employer or his agents of either implied or express terms of the contract, e.g. if the Employer in some way hindered progress of the Works or if the architect were negligent in carrying out his duties, resulting in loss to the contractor.

Ex gratia claims

These have no legal basis but are claims which the contractor considers the Employer has a moral duty to meet, e.g. if he has seriously underpriced an item whose quantity has been increased substantially because of a variation or the remeasurement of a provisional item, which will in consequence cause him considerable loss. The Employer is under no obligation to meet such 'hardship claims' but may be prepared to do so on grounds of natural justice or to help the contractor where otherwise he might be forced into liquidation.

Contractual claims

These arise from express terms of a contract and form by far the most frequent kind of claim. They may relate to any or all of the following:

(*a*) fluctuations;
(*b*) variations;
(*c*) extensions of time;
(*d*) loss and/or expense due to matters affecting regular progress of the works.

Fluctuations claims

These relate to increases in the costs of labour, materials and plant and to levies, contributions and taxes, which the contract provides for the contractor to be reimbursed by the traditional method in clauses 38 and 39 of JCT 80. This has been fully explained in Chapter 9 and requires no further treatment here.

Claims arising from variations

These may relate to one or more of the following clauses:

(i) clause 13.5.1.2 – the surveyor may have priced at bill rates variation items which are apparently similar to bill items but which were not executed under similar conditions to those envisaged at the time of tendering. For example, an Architect's Instruction changing the kind of facing bricks may affect the time taken for laying the bricks, and may lead to more sorting if there is a higher proportion of misshapen bricks or if they are more susceptible to damage in handling. The contractor would be entitled to an increase over the bill rates.
(ii) clause 13.5.3.3 – the contractor may claim that adjustment of one or more Preliminary items should be made, though the surveyor has not included the adjustment in his valuation of a variation. For example, an Architect's Instruction changing the pointing of faced brickwork from 'pointing

as the work proceeds' to 'pointing on completion' may require scaffolding to be left standing for a longer period than would otherwise have been necessary. If scaffolding is priced in the Preliminaries (as is usual), the relevant item should be adjusted.

(iii)   clause 13.5.5 – it may be claimed that, in consequence of a variation, the conditions under which other work is carried out has changed and therefore the bill rates for that work should be adjusted to reflect the changed conditions. Thus, a variation changing part of the foundations of a building from concrete deep strip to short bored piles would, it might be claimed, so reduce the total volume of concrete required for the rest of the deep strip foundations that the bill rates for the concrete would no longer apply. If this argument can be substantiated, the rates should be adjusted.

(iv)   clause 13.5.6 – any of the bases of claim described in the preceding three paragraphs may also apply to Performance Specified Work to which this clause relates.

(v)   clause 13.5.7 – the contractor may claim that he has incurred additional costs which are directly associated with a variation but for which he has not been reimbursed by the application of the normal valuation rules. An example of such a claim is given on p. 117.

Claims for extensions of time

These arise from clause 25 of JCT 80. Clause 23 requires the contractor to complete the Works on or before the Date for Completion stated in the Appendix (see p. 347) or such later date as may be fixed by the architect under clauses 13A or 25. If the contractor fails to do so, he becomes liable for liquidated damages (clause 24), which the Employer is entitled to deduct from payments due to the contractor at the rate stated in the Appendix for the period between the date when completion should have taken place and when it actually took place. This action of the Employer is subject to the prior issue by the architect of a certificate of non-completion by the Completion Date.

The rate of liquidated damages stated in the Appendix should be a genuine pre-estimate of the likely loss to the Employer due to the time overrun and is not adjustable according to the actual loss incurred. The purpose of granting extension(s) of time is only to relieve the contractor of liability to pay liquidated damages for the period of the extension and does not carry an automatic right to reimbursement of any loss or expense which the contractor may claim he has suffered or incurred due to the matter for which the extension was granted.

Clause 25.4 lists Relevant Events, the happening of any one of which is a ground for extending the contract time by such period(s) of time as the architect estimates is 'fair and reasonable', if it is 'apparent that the progress of the Works is being or is likely to be delayed'. The contractor must give prompt written notice of such actual or likely delay, (i) stating the cause(s), (ii) identifying the Relevant Events, (iii) giving particulars of the expected effects, and (iv) stating the estimated extent of the delay in the completion of the Works. The architect is under no obligation to act until such notice has been received.

Within 12 weeks of receipt of all the above information, the architect must give a fair and reasonable extension of time if he is of the opinion that the cause of delay is a Relevant Event and that completion of the Works is likely to be delayed thereby. If an Architect's Instruction calls for the omission of work, the architect may fix an earlier completion date than that fixed on a previous occasion, provided it is not earlier than the Date for Completion stated in the Appendix.

After the Completion Date, if that occurs before actual completion, the architect must review his previous decisions (if any) on extensions, not later than the end of 12 weeks after the date of actual completion. He must then (i) fix a later Completion Date, (ii) fix an earlier Completion Date, or (iii) confirm the Completion Date previously fixed and notify his decision to the contractor in writing.

Claims for loss and/or expense due to matters affecting regular progress

These claims arise under clauses 26 or 34 of JCT 80. Clause 26.2 lists eight matters (in addition to deferment of giving possession of the site to the contractor) which might materially affect progress. If the contractor believes that any one or more of those matters has affected the regular progress of the Works and that he has, in consequence, incurred loss and/or expense, then he may apply in writing to the architect for reimbursement. If the architect is of the opinion that such loss and expense has been or is likely to be incurred due to one or more of the stated matters, he must ascertain or (as normally happens) he must ask the quantity surveyor to ascertain the amount of such loss and expense which has been or is being incurred.

All the matters listed in clause 26.2 are those for which the Employer is responsible, being acts or omissions of the Employer or the architect as his agent. Excluded from recovery are all other causes of loss or expense, such as those which are unknown or undefined or the result of commercial risks, bad pricing of the contractor's tender, bad weather or bad organization or management of the contract. Nor is there any automatic right of recovery of loss solely because an extension of time has been granted. On the other hand, it is possible for the contractor to have suffered recoverable loss or expense when no extension of time has been granted.

The architect need take no action unless and until a written application which clearly states the circumstances which have caused the loss has been received from the contractor. A telephone request or an oral application made to the architect on the site is insufficient; also a new application is necessary for each new matter that arises. If the architect fails to act upon receipt of an application properly made, the Employer will become in breach of contract and liable for damages.

The contractor is not required to formulate a claim, but must provide all necessary information to enable the amount of the loss or expense to be ascertained. This means that the contractor must be prepared to reveal what he probably considers to be confidential information about such matters as actual wages and bonuses paid to operatives, details of head office overheads, etc. 'Ascertain' means 'to find out exactly', so it is necessary for the quantity surveyor to find out the actual loss and not to rely upon estimates or the use of formulae in order to arrive at a sum.

The loss and/or expense, to be recoverable, must have been a direct consequence of the matter(s) referred to in the claim without there having been any intervening cause. Thus, if the Employer fails to provide materials for the Works which he has agreed to provide and the contractor is asked to obtain them, then the contractor is entitled to recover the expense incurred in so doing, as well as the loss due to delay in obtaining them. If, however, in obtaining those supplies, deliveries are held up because of a strike in the haulage industry and additional loss or expense ensues, this would not be recoverable, being an indirect result of the original cause.

If the contractor does formulate a claim or even if he just provides required information, he is much more likely to succeed in his claim if he makes a clear and orderly presentation of the data. It will not help the architect in forming an opinion as to the soundness of a claim or the surveyor in ascertaining the proper amount if they are presented with a jumbled mass of papers which it is virtually impossible to sort out.

Heads of claim may be any or all of the following:

(i)   inefficient use of labour and/or plant;
(ii)   increases in costs of labour, materials, etc. during the period of disruption;
(iii)   site running costs;
(iv)   head office overheads;
(v)   finance charges and interest;
(vi)   loss of profit.

(i)   Inefficient use of labour and/or plant refers to men and plant standing idle or working at a reduced level of output. It would be necessary for the contractor in substantiating a claim to produce evidence of the estimated levels of output used in preparing his tender and records of the actual output during the disruption period.

(ii)   Increases in costs would normally only apply in the case of firm price contracts, although they could possibly apply in the case of fluctuations contracts using clauses 38 or 39, in respect of materials, etc. not listed on the list of basic prices of materials.

(iii)   Site running costs relate to site staff costs, offices, messrooms, sheds, rates on temporary buildings, etc., normally referred to as Preliminaries. Prolongation of a contract could result in some or all of these items being required for a period in excess of that for which the contractor allowed in his tender.

(iv)   Head office overheads include the cost of maintaining head and branch offices, plant and materials yards, rents, rates,

directors' and staff salaries, office running expenses, travel-
ling expenses, professional fees and depreciation. Because of
the complexity of trying to sort out such costs, a formula is
often used as a means of evaluation. Three in common use
are Hudson's, Emden's and the Eichleay formula.[1] The first
two are similar:

$$\frac{h}{100} \times \frac{c}{cp} \times pd$$

where h = head office overheads and profit per cent included
in the contract (Hudson's), or h = per cent arrived at by
dividing total overhead costs and profit of the contractor's
organization as a whole by total turnover (Emden's); c =
contract sum; cp = contract period in weeks; and pd = period
of delay in weeks.

Many surveyors do not accept the use of formulae on the
grounds that actual cost only is admissible and that the
contractor must specify precisely what actual additional
expense has been incurred.[2] The decision in *Tate & Lyle
Food Distribution Co Ltd v Greater London Council (1982) 1
WLR 149* supports this view.

However, in the case of *J E Finnegan Ltd v Sheffield City
Council (1988) 43 BLR 124*, it was held that the Hudson
formula should be used to calculate head office overheads
and profit as part of the plaintiff's claim. However, the
learned judge, in referring to it, was confusing the Hudson
formula with the Emden formula. But it remains the law
that the contractor must establish proof of actual loss, even
when one of the formulae is used, which should only be by
agreement between the Employer and the contractor.

(v) Finance charges or interest – finance charges are interest
charges incurred by the contractor on money he has had to
borrow (or interest that he was prevented from earning on
his own capital) in order to finance the direct loss and/or
expense claimed. The rates of such interest must be those
actually paid to the bank or other finance source and
certified by the contractor's auditors. Interest due to late
payment of sums owing (such as retention monies) is not
admissible.

(vi) Loss of profit refers to profit which the contractor could have earned but was prevented from earning as a direct result of one or more of the matters listed in clause 26.2. This could arise where the architect omitted a substantial part of the project, say one wing of a new hospital. In such a case, the contractor would be entitled to reimbursement of the profit which he could prove he would have made had he been allowed to complete the contract as originally envisaged. It might also happen that because of prolongation of one contract, the contractor was prevented from taking on another. If he could prove his loss of profit as arising directly from a clause 26 matter, he would be entitled to reimbursement.

It should be noted that acceleration costs, i.e. additional costs incurred due to speeding up progress of a contract in order to meet the Completion Date, are not recoverable, nor are the costs of preparing a claim or providing detailed information to enable loss or expense to be ascertained.[3]

Having ascertained 'from time to time' amounts of loss and expense incurred, the architect is required to include them in the next interim certificate after each ascertainment. Such amounts should be paid in full, i.e. they are not subject to retention. They will also be included in the final account for adjustment of the contract sum.

The provisions of clause 34 in regard to loss and/or expense incurred as a result of disruption following the finding of antiquities, fossils, etc. are similar to the provisions in clause 26. The main difference is that the contractor is not required to give written notice to the architect as a pre-condition to consideration of a claim, he is merely required to inform him and to await his instructions.

References

1. KNOWLES, R., Calculating office overheads, *Chartered Quantity Surveyor*, 1985, Vol. 8, No. 5, p. 207.
2. MORLEDGE, R., Letter, *Chartered Quantity Surveyor*, 1986, Vol. 8, No. 7, p. 5.
3. RICS, *Contractor's Direct Loss and/or Expense* (London: RICS Books, 1987), p. 11.

Bibliography

1. POWELL-SMITH, Vincent, *Problems in Construction Claims* (London: Blackwell Science, 1990).
2. TRICKEY, G., *Presentation and Settlement of Contractors' Claims* (London: Spon, 2nd edn, 1996).
3. NEWMAN, P., *Loss and Expense Claims Explained* (London: RIBA Publications, 1994).
4. KNOWLES, R., *Claims: Their Mysteries Unravelled* (London: Knowles Publications, 2nd edn, 1993).
5. POWELL-SMITH, Vincent, SIMS, John and CHAPPELL, David, *Building Contract Claims* (London: Blackwell Science, 1996).

12

Interim valuations

When the value of contracts is more than a few thousand pounds, it is normal practice for contractors to be paid sums on account as the construction work proceeds. This is because it is generally recognized that it would be unreasonable to expect contractors to finance construction operations without assistance from Employers. Also, the expense of borrowing large sums, which otherwise would be involved, would add significantly to total costs – which it is in the interest of Employers to minimize. All the Standard Forms include provisions for periodic or 'interim' payments to be made for these reasons.

It is universally accepted that interim payments are approximate only and, provided the amounts included for the various constituents are reasonable, no objection will be raised because they are not exact. For this reason, the amounts shown in certificates are always round pounds, as also are the sums in the main money column of valuations.

Frequency of valuations and 'valuation date'

Clause 30 of the JCT Form provides for the Employer to pay the contractor such sums as are stated to be due in Interim Certificates issued by the architect at the periods stated in the Appendix to the Conditions of Contract. In practice, this normally means monthly, commencing one month after the date on which the contractor takes possession of the site. The architect usually relies on the quantity surveyor to advise the sum which should be stated as due in any such certificate. In order to do this, the surveyor prepares a valuation in accordance with the provisions of clause 30 of the JCT Form.

As an alternative to regular monthly certificates, the contract conditions may provide for 'stage payments',[1] i.e. for payments to be made when certain defined stages in the construction work have been completed. Thus, the first payment may be due when the substructure is finished and ground slabs laid, the second when superstructure walls and upper floors are constructed, the third when the roof is finished, and so on. Stage payments are more appropriate to housing contracts than to more complex projects where the stages are often not so readily or satisfactorily definable. Table 12.1 shows a breakdown of a housing contract, consisting of four types of house, into stages in preparation for interim valuations.

Table 12.1 Typical breakdown of house type contract sums for use in 'stage payments' method of interim valuations

| Stages | Proportion of contract value for house types | | | | Totals |
	A	B	C	D	
	£	£	£	£	£
Substructure	2 300	2 480	2 392	2 852	10 024
External walls, upper floors, windows and external doors	6 780	7 868	7 164	8 856	30 668
Roof, internal walls and partitions, first fixings	5 874	6 720	6 256	7 954	26 804
Plumbing, internal finishings, second fixings	7 486	8 332	7 904	8 432	32 154
External works	2 480	2 912	2 770	4 360	12 522
Preliminaries	3 700	3 964	3 838	4 704	16 206
Totals	28 620	32 276	30 324	37 158	128 378

As the sum eventually certified as due for payment relates to work actually completed and unfixed materials and goods actually stored on the site, it is necessary for the surveyor personally to visit the site to see for himself both the completed work and the unfixed materials and goods and then to ascertain their total value.

Where the contract provides for advances at regular intervals, it will be advantageous to decide from the start a fixed date in each month on which to do valuations. Although the surveyor has no obligation under the terms of the JCT Form to notify the contractor when he proposes to visit the site for valuation purposes, nevertheless it is desirable as well as courteous to do so and even better if the surveyor and the contractor's surveyor agree on a 'valuation date' in each month.

From the contractor's point of view, the exact position of the day in the month may well affect his cash flow as, for example, his invoices for materials become payable at the end of the month. Also, if the formula method of recovery of fluctuations applies (see p. 155), the fluctuations amount included in a valuation will be affected by the relationship between the 'valuation date' and the date of publication of the monthly indices.

From the Employer's point of view the 'valuation date' will affect his cash flow situation also. If the Employer is a public authority and it is necessary for payments to be approved by a finance committee, then the date of the committee meeting in each month may have a direct bearing upon when the valuation ought to be done, if payment is not to be delayed unduly. The surveyor should give careful consideration to any such factors involved, including his own commitments in regard to other contracts already running.

General procedure

The proper way to prepare an interim valuation is to value, on each occasion, the amount of work which has been done since the beginning of the contract and the value of unfixed materials and goods on the site on the 'valuation date'. From the total value so arrived at, the total of previous payments on account, if any, will be deducted, leaving a balance due for payment. By following this procedure strictly, any under-valuation or over-valuation of either work completed or of unfixed materials on the last previous 'valuation date' will be automatically corrected. The surveyor should not attempt to value in isolation the work done and the materials delivered since the last valuation.

Experience will confirm that valuations are most conveniently set out on double billing paper, thus allowing for sub-totals to each section and subdivision being carried over into the right-hand money column (*see* example on p. 195). Such sub-totals are often useful for carrying forward into subsequent valuations.

Preparing the valuation on site

The surveyor's first task on visiting the site for valuation purposes (having first made his presence known to the site agent), is to tour the Works, making notes as necessary of the extent of work done and listing the quantities of the various materials and goods stored on the site. He will then be in a position to get down to preparing a draft valuation in the site agent's or clerk of works' office (if there is no office for the surveyor's exclusive use, as often there is not). It is highly desirable, if at all possible, that the valuation total be determined before leaving the site.

As already stated, the surveyor and contractor's surveyor will usually agree to meet on site at a mutually convenient time on the 'valuation date' in the month and will prepare the valuation together, although, of course, the responsibility for the resulting recommendation to the architect will be that of the surveyor alone. Thus, the total sum and its constituent amounts will normally be agreed before leaving the site and subsequent dispute will be avoided. Sometimes, the contractor's surveyor is happy to leave valuations entirely to the surveyor and will accept whatever amounts the architect certifies.

Some contractors prefer to prepare detailed applications for interim payments themselves which they submit to the surveyor a few days before the valuation date. The surveyor has no obligation, however, to make use of them, although he may find it convenient to use them as a basis for his own valuation by checking each of the constituent amounts and, if necessary, amending any as he sees fit.

Even when such detailed applications are not made, it may be very helpful to have a list, prepared beforehand by the site agent, of the quantities of unfixed materials stored on site. Surveyors often arrange with contractors for such a list to be available when they arrive on site to prepare the valuation. The surveyor should then check the list, including the quantities, when making the initial tour of the Works.

Inclusions in valuations

Clause 30.2 of the JCT Form provides that 'the amount stated as due in an Interim Certificate ... shall be the gross valuation' of specified constituent parts of the Works, less the Retention Percentage and the total of previous interim certificates. In order to determine 'the gross valuation', the value of each of the constituent parts has to be ascertained. Those parts may be any or all of the following:

1 Preliminaries.
2 Main contractor's work (as billed).
3 Variations.
4 Unfixed materials and goods.
5 Statutory fees and charges.
6 Nominated sub-contractors' work.
7 Nominated suppliers' goods.
8 Fluctuations in costs of labour, materials and/or taxes, etc.
9 Retention.

Each of the items on the above list will now be considered in detail. The examples at the end of the chapter illustrate the application of most of the principles and procedures discussed.

1. Preliminaries

Where there is a bill of quantities, the evaluation of the Preliminaries will usually be less difficult than where there is not. In the latter case, it will be necessary, before the date of the first valuation, to make an apportionment of the contract sum to each of its principal constituents. The Preliminaries constituent will then need to be broken down into separate amounts such as would normally be seen in the Preliminaries section of a bill of quantities. This process will be very much easier and the results be more satisfactory, if there is a Contract Sum Analysis (see p. 73) which shows the total of Preliminaries included in the tender. The contractor will be able to produce the build-up of the tender and the individual amounts allowed in it for the main Preliminaries items. These amounts will then be available for valuation purposes – and later, for the adjustment of Preliminaries, if necessary, in the final account – just as if there had been a bill of quantities.

Even where there is a bill, the situation may still present difficulties. The contractor may have put amounts against all, some, one or none of the priceable items in the Preliminaries section, having included the balance of their value, if any, in other parts of the bill. The surveyor can only take account, of course, of those items which have been priced, even though they may represent only part of the total cost of the Preliminaries. It will help in arriving at realistic valuations if the sums which have been inserted in the Preliminaries section are analysed, their constituent parts being dealt with as described in the following paragraphs. Again, such analyses will be more satisfactory if provided by or agreed with the contractor, preferably before the contract is signed.

Preliminaries items (or their components) are of four kinds, namely, cost-related, time-related, single-payment or a combination of two or more of the others. Those which are cost-related (e.g. Water for the Works) depend for their value on that of the remainder of the contract sum or of its labour content. Time-related items (e.g. site supervision) depend for their value on the contract period. Single-payment items (e.g. the provision of temporary access roads) are those whose value is not affected either by the value of the rest of the contract or by the contract period but are carried out at a particular point in the progress of the Works. The cost of some Preliminaries items consists of one or two single-payment components and a time-related one, e.g. the provision of a tower crane, involving single payments for erection and dismantling and a weekly hire charge for the intervening period.

The surveyor's next task prior to the first valuation, after having ascertained, where appropriate, the breakdown of the Preliminaries items which have been priced, is to put them into the categories described above, so that the total of each category is then readily available for each valuation thereafter. Table 12.2 illustrates how this may be done. See p. 120 with regard to the rules in SMM7 for the pricing of Preliminaries items which will facilitate their breakdown into the categories shown in Table 12.2.

Table 12.2 Three-storey block of shops and flats, Thames Street, Skinton. Breakdown of Preliminaries items*

Items	Time-related	Fixed charges (a)	Fixed charges (b)	Cost-related	Totals
	£	£	£	£	£
Management and staff	20 000	–	–	–	20 000
Site accommodation	720	270	90	–	1 080
Electric lighting and power	180	–	–	–	180
Water	–	117	–	1 260	1 377
Telephones	450	–	–	–	450
Safety, health and welfare	2 880	450	270	–	3 600
Cleaning	–	–	540	–	540
Drying out	–	–	450	–	450
Security	1 350	–	–	–	1 350
Small plant and tools	–	100	250	2 700	3 050
Setting out	–	540	–	–	540
Temporary roads	–	1 170	180	–	1 350
Scaffolding	900	450	100	–	1 450
Fencing and hoardings	270	1 080	270	–	1 620
Insurance against injury	–	–	–	3 632	3 632
All risks insurance	–	–	–	811	811
	26 750	4 177	2 150	8 403	41 480

Notes: Total of time-related amounts = £1337 per month for the Contract Period of 20 months.

Fixed charges (a) are sums expended at or soon after the commencement of the Contract for erection, installing, etc.

Fixed charges (b) are sums expended mainly towards the end of the Contract for dismantling, removal, etc.

Total of cost-related amounts = 2.5 per cent of the Contract Sum less p.c. sums and Preliminaries.

* See Appendix A for full details of the Contract.

It is not uncommon to deal with the total value of the Preliminaries as if the items were either all cost-related or all time-related.[2] Thus, in the first case, the total of the Preliminaries is calculated as a percentage of the contract sum after deducting the Preliminaries and the prime cost sums. In each valuation, this percentage is applied to the total value of item 2 of the list on p. 181. In the second case, the total of the Preliminaries is divided by the contract period (in months) and the resulting sum is multiplied by the number of months which have elapsed to date. This amount is then included in the valuation as the total amount for Preliminaries.

There is no practical reason why the Preliminaries should not be dealt with in either of these two ways, although it should be recognized that they give only approximate results. It is argued in their defence that these methods save time, although it should be pointed out that once the allocation of amounts to the four categories as illustrated in Table 12.2 has been done, it takes very little longer each month to calculate more nearly the true value.

The objection to both these methods is that they result in overpayment of some items and and underpayment of others. Also, where the time-related method is used and the contract runs behind schedule, there will be a danger of overpayment, unless the fixed monthly amounts are adjusted. Where the cost-related method is used, there is a danger of inadvertently exceeding the total of the Preliminaries when the total value of the contractor's own work has been significantly increased by variations.

None of the foregoing objections matters very much when the contract is keeping to schedule and no real problems arise. They will matter, however, if the employment of the contractor is determined because of insolvency. Then, the likelihood will be that the Preliminaries will have been overpaid, particularly if the contract is in its early stages. The possibility of such embarrassment to the surveyor (to put it at its best) will be avoided if the small amount of extra time and trouble is taken to ascertain more exactly the value of the Preliminaries.

2. Main contractor's work

The value of the work carried out by the contractor's own workforce will be readily ascertainable from the measured work sections of the bills of quantities or, in the case of lump sum contracts without quantities, from the Contract Sum Analysis, if provided, or from a schedule drawn up on the basis of an analysis of the tender. The prices contained in the bills of quantities, being 'the Contract Bills',[3] must be used, of course, in the valuation of the main contractor's work, regardless of whether the contractor under-priced or over-priced the work when compiling the tender.

The surveyor, when valuing the main contractor's work, will begin with the 'Groundwork' section (or the 'Demolition/Alteration/Renovation' section if there is one) and proceed in order through all the succeeding work sections which contain items of work which have been wholly or partly carried out. He should put down separately the total valuation of such work within each work section or sub-section, indicating by use of bill item references what is included in each amount. The reason for so doing is that every amount can then be verified and substantiated subsequently, should any query arise. Also, such detail will often prove useful when doing the next valuation. Of course, if the whole of a work section has been completed, it is only necessary to show its total. The examples of valuations given at the end of the chapter illustrate the procedure.

The notes made during the initial tour and inspection of the Works will again be used in assessing approximately the value of any partly completed items, sections or sub-sections. The completed work as a percentage of the whole item should be indicated opposite the bill reference. Actual measurement of the work on site will seldom, if ever, be necessary solely for interim valuation purposes, except in the case of contracts not based on bills of quantities.

3. Variations

Very few projects, if any, have no variations. On some jobs they number in double or even treble figures. They present a difficulty in the context of interim valuations for several reasons. First, clause 30.2.1.1 of the JCT Form requires that effect be given in Interim Certificates to 'the measurement and valuation' of variations. For this to be done, each variation must be valued, either approximately or accurately, as soon as possible after issue. Pressure of other work may cause delay in dealing with variations but it may be dangerous to have to resort to guesswork in consequence. It is wise, therefore, to give priority to them, even if only to the extent of an approximate evaluation of the net effect of each variation, before each valuation date. The methods of valuing variations are discussed in Chapter 8.

Secondly, the question of how best to deal with the effect of omissions on the measured sections of the bills of quantities is bound to arise. There are two alternatives. The first is to ignore the variations when dealing with the work measured in the bills, the omissions being allowed for when adding the net value of variations. The second is to take the omissions into account when valuing 'main contractor's work' and adding the value of additions only against the 'variations' subhead.

In practice, a combination of the two alternatives will serve best. The first may be used as a general rule, the second being used where a complete section or group of items in the bills is affected by a variation. For example, a variation increasing the width of some (but not all) of the windows will affect part of the items for lintels, cavity trays, sills, window boards, etc. In such a case, it will probably be simpler to include the full value of the items as originally billed in the value of 'main contractor's work', leaving the net value of the variation to take care of the omissions.

If, however, a variation is ordered changing the specification of all the internal doors from painted plywood-faced to hardwood-veneered and the linings and architraves from softwood to hardwood, then the second alternative will probably be preferable, that is, to exclude the whole of the appropriate sub-sections of the Windows/Doors/Stairs and Building Fabric Sundries sections of the bills from the value of 'main contractor's work' and include in the 'variations' part of the valuation the total value of the additions items.

4. Unfixed materials and goods

Clause 30.2.1.2 of the JCT Form requires the inclusion in the amount of an interim certificate of materials and goods delivered to the site for incorporation in the Works. In order to value them, the surveyor will need to be satisfied that the materials and goods are actually on the site and to ascertain approximately how much of each material or good there is. It is not necessary to know the exact quantities because by the time the next valuation is done, most or all of the materials will have been incorporated into finished work which will then be valued as such.

It is helpful if the contractor's surveyor prepares a list of all the materials with their respective quantities, immediately prior to the valuation date and has supporting delivery notes and invoices available so that each of the materials can be valued at invoice cost. The surveyor will have then only to check that the materials and goods are on site and that the quantities and value are as stated. If not, he will amend the list accordingly. The most recent deliveries will probably not have been invoiced as yet and the costs of the equivalent types and qualities on the latest available invoices will be used instead.

It will be necessary to reduce or exclude the invoice value of (*a*) insufficiently protected materials or those which have deteriorated or been damaged; (*b*) quantities which are clearly in excess of requirements; and (*c*) any materials or goods which have been delivered prematurely, that is, when the length of time before they are likely to be required is unreasonably long.

The JCT Form includes a provision in clause 30.3 for the value of materials and goods intended for the Works but not yet delivered to the site to be included in the certificate amount at the discretion of the architect, subject to a number of extra conditions. Briefly, these are: (*a*) that they are intended for incorporation into the Works and are in accordance with the contract, and that they need only be transported to the site to be ready for use; (*b*) that they are set apart at the place where they are being kept and are clearly identifiable and appropriately marked; (*c*) that ownership is established in written contracts or sub-contracts for their supply; and (*d*) that they are properly and adequately insured. The surveyor must satisfy himself that all these conditions have been met before he accepts a claim for the value of the goods to be included in a valuation.

The question whether to allow for overheads and profit on the value of unfixed materials and goods sometimes arises. This may at first appear to be a reasonable suggestion, as the value of each part of the Works is presumed to include those factors and the materials and goods are to become part of the Works. It may appear also as a reasonable interpretation of the words 'total value' in clause 30.2.1.2. The Q.S. (Practice and Management) Committee of the Royal Institution of Chartered Surveyors has expressed the opinion that 'the word "total" is used only to indicate the collective value of work executed and materials delivered . . .' and 'could not be interpreted as including anything in addition to their value *for the purpose of the interim certificate*'.[4]

5. Statutory fees and charges

Any sums paid by the contractor to any local authority or statutory undertaker in respect of fees or charges for work executed or materials or goods supplied in the course of carrying out its statutory obligations should be included in the next interim valuation thereafter. Such charges would commonly be for connections to water, gas, and electricity mains.

Normally, there will be provisional sums included in the bills of quantities in respect of such work. It should be noted that clause 6.3 of the JCT Form specifically excludes the operation of clause 35 to work done by statutory bodies. In consequence, there will be no question of adding any sums equivalent to cash discounts to the invoiced amounts, which invariably will be net. This exclusion only applies to work done or goods supplied in discharge of the statutory obligations of the authorities. It does not apply to such work or goods which are outside the scope of those obligations, such as where they have been tendered for in competition with other sub-contractors and/or suppliers.

6. *Nominated sub-contractors' work*

Clause 4.17 of the JCT Nominated Sub-Contract Conditions (NSC/C) spells out in detail what should be included in respect of sub-contract works. It is common practice for nominated sub-contractors to make application from time to time, through the main contractor, for the amounts on account to which they consider they are entitled.

It will be to the advantage of all concerned, therefore, if nominated sub-contractors were informed at the commencement of the contract, or as soon thereafter as reasonably possible, of the valuation date in each month and asked to make applications for payments on account in time for inclusion in the next valuation. Then, in each case, if the surveyor is satisfied that work to the value stated has been carried out and materials to the value stated are stored on the site, he can include the amount applied for in his calculations. Profit and attendance additions should be made *pro rata* to the appropriate items in the bills of quantities, as explained in Chapter 7. Of course, if the work of any sub-contractor were complete, the full amount of attendances would be included. As in the valuation of main contractor's work, the sums included for nominated sub-contractor's work should be the total value of the whole of their work carried out to date, not for work done during the preceding month only.

Clause 35.13.3 of the JCT Form requires the contractor to provide proof of payment to any nominated sub-contractor of any sum included in the immediately preceding certificate in respect of their work, materials or goods. The standard valuation form (see p. 196) published by the Royal Institution of Chartered Surveyors has a footnote saying 'It is assumed that the Architect/The Contract Administrator will ... satisfy himself that previous payments directed for Nominated Sub-Contractors have been discharged'. Nevertheless, it is considered to be good practice to agree with the contractor, at the first meeting for valuation purposes, that he will produce, on each valuation date, proof of payment of all such amounts previously certified.[5] Should such proof not be forthcoming in respect of any sum the surveyor should inform the Architect/C.A. accordingly, so that he may proceed in the manner prescribed in clause 35.13.5 of the JCT Form.

Notwithstanding the absence of such proof, if the architect is satisfied that the contractor has properly discharged all amounts to a sub-contractor that have been included in previous certificates, then the prescribed default procedure will not be implemented (*see* clause 35.13.4 of the JCT Form).

If for that (or some other) reason, the Employer has paid a nominated sub-contractor direct, the contractor will still be entitled to the appropriate sums for profit and attendance. He will not be entitled, however, to any sum equal to the cash discount, as (*a*) he has not paid the sub-contractor, and (*b*) the entitlement has been lost because of failure to pay within the period stipulated in clause 4.16.1.1 of the JCT Nominated Sub-Contract Conditions (NSC/C). In the preparation of subsequent valuations, the total value of the sub-contractor's work should be shown in the normal way, plus profit and attendances. The deduction of the amount paid direct should be made, not by the surveyor or the architect, but by the Employer.

Payments on account to nominated sub-contractors should include the value of unfixed materials on the same conditions as apply to the main contractor's materials (*see* clause 4.17 of the JCT Nominated Sub-Contract Conditions (NSC/C). Another inclusion may be payments in respect of fluctuations. Whether a sub-contractor is entitled to fluctuations payments depends upon the terms of the sub-contract. If he is, then the calculation of the sums due will normally (but not necessarily)[6] be made in the same manner as is applicable to such sums due to the main contractor in respect of his own work. Profit and discount amounts should be calculated on the *total* payments to sub-contractors, including any fluctuations.

7. Nominated suppliers' goods

The value of goods and/or materials supplied by nominated suppliers should be included in valuations on the same basis as those supplied by non-nominated suppliers, namely, that the goods or materials are properly protected and stored whilst still unfixed and are not damaged or defective. Amounts should be inclusive of the proper discount for cash,[7] also for packing, carriage and delivery. Amounts should be exclusive of any trade discount allowed.

Where an invoice includes increases over the sum(s) stated in the quotation on which the order was based, whether such increases are admissible or not will depend on (*a*) the cause, and (*b*) the conditions of the contract of sale as stated in the quotation. If the cause is the imposition or increase of a tax or duty, the appropriate amount of the increase is admissible.[8] If the increase arises from some other cause, it will normally only be acceptable if the quotation specifically included a provision for such increased charges to be added.

8. Fluctuations in costs of labour, materials, etc

The amounts, if any, under this heading which should be included in interim valuations, will vary according to whether the contract is a firm price or fluctuating price one (see p. 140) for definitions of these terms). The subject of fluctuations has been dealt with in detail in Chapters 9 and 10 and it will suffice to add only two further points here. First, where the 'traditional' method of recovery is allowed for in the contract conditions, amounts due should be included in interim valuations as early as possible after the costs have been incurred. Secondly, the amounts of fluctuations calculated by that method are payable in full (that is, they are not subject to retention) with the percentage addition (if any) referred to in clause 39.8 of the JCT Form and as stated in the Appendix to the Conditions of the contract. It should be noted, however, that amounts calculated by the 'Formula Rules' method are subject to retention (*see* clause 30.2.1.1 of the JCT Form).

9. Retention

It is a common provision in all the standard forms of building contract for a percentage of the valuation total to be deducted. Thus, the sum of money deducted is said to be 'retained' by the Employer and is called 'the Retention' in the JCT Form. The percentage used to calculate the amount of the Retention is called 'the Retention Percentage' (see clause 30.4.1.1).

The purpose in retaining part of the total value of work completed to date is (*a*) to provide an incentive for the contractor to complete the Works promptly, and (*b*) to 'cushion' the Employer to some extent against the effects of the contractor defaulting, should that happen.

The actual percentage retained will be that stated in the contract Conditions. Clause 30.4.1.1 and the Appendix to the JCT Form (*see* p. 348) provide for 5 per cent unless a lower rate is specified, and in a footnote to clause 30.4.1.1, the document says 'Where the Employer at the tender stage estimates the Contract Sum to be £500,000 or over, the Retention Percentage should not be more than 3 per cent'.

The exclusions from the application of the Retention Percentage where the JCT Form applies are set out in clause 30.2.2. For convenience of reference, they are listed as follows:

(*a*) statutory fees and charges (clause 6.2);
(*b*) the cost of correcting any errors in setting out the Works, for which the contractor is not held responsible (clause 7);
(*c*) the cost of opening up work which has been covered up and/or the cost of testing materials, goods or executed work (clause 8.3);
(*d*) costs relating to patent rights (clause 9.2);
(*e*) the cost of making good faulty work for which the contractor is not held responsible (clauses 17.2 and 17.3);
(*f*) insurance premiums payable by the contractor in order to maintain insurances against damage to property other than the Works (clause 21.2.3);
(*g*) insurance premiums which should have been paid by the Employer but in regard to which he has defaulted (clauses 22B and 22C);
(*h*) loss or expense due to the regular progress of the Works having been affected by specified matters (clauses 26.1 and 34.3);

(*i*) final payments to nominated sub-contractors (clause 35.17);
(*j*) fluctuations in the cost of labour and materials, etc., calcu-
       lated other than by the 'Formula Rules' method (clauses 38
       and 39).

The JCT Form provides in clause 30.4 for one-half of the total
then retained to be paid to the contractor in the next interim
certificate after the architect has issued a certificate of Practical
Completion, subject, of course, to any releases of retention
already made. Footnote [t] on p. 53 of the JCT Form makes this
clear.

The same provision will operate under clause 18 of the JCT
Form, should the Employer take possession of any part of the
Works before the Date for Completion stated in the Appendix to
the Conditions of the contract. In that circumstance, the amount
of retention monies released will be one-half of that proportion of
the total retention which the estimated value of 'the relevant
part' bears to the total value of the Works.

The second half of the retention monies is released when the
Defects Liability Period stated in the Appendix to the JCT Form
has expired and the contractor has made good, at the architect's
request, such defects as have appeared during that time. A
similar provision applies to any part of the Works of which the
Employer has taken early possession.

The architect may, if he so wishes, subject to certain conditions,
release the whole of that part of the retention monies which
relates to any nominated sub-contractor's work, at any time after
Practical Completion of the sub-contract work. This action
becomes obligatory upon the expiry of twelve months from the
certified date of Practical Completion of the sub-contract work.
The reason for this is that it is unreasonable that a nominated
sub-contractor who completes his work early in the contract
period should have to wait many months for the payment of the
balance due to him.

It will be seen from the above, that it is possible, at a particular
'valuation date', for the value of part of the Contract Works to be
subject to the full retention percentage, that of another part to be
subject to half that percentage, and the rest not to be subject to
any retention at all. The surveyor must be careful to ensure that
the correct retention is deducted from those parts of the total
value, and only those parts, to which it properly applies.

Sometimes, contractors object to retention being deducted from the value of goods and services for which they are obliged to pay in full, for example, materials and goods supplied by nominated suppliers (clause 36.4.6 of the JCT Form allows deduction of cash discount only). Whilst this situation may appear to be unfair to contractors, the proper action under the JCT Form is for the surveyor to deduct retention from the whole of a valuation, less only amounts coming within the exclusions listed on pp. 191–2.

Notifying the architect

When the surveyor has completed his valuation, he must inform the architect as soon as possible of the amount of the payment recommended. Prompt notification is essential if the architect's certificate is to be issued within the period from the valuation date prescribed in clause 30.2 of the JCT Form, that is, seven days. The Royal Institution of Chartered Surveyors publishes a standard form for this purpose in sets of three forms (one copy each for architect, contractor, and surveyor) in pads of 100 sets. The form is shown in Figure 12.2, completed to correspond with the first of the examples of draft valuations.

The form provides for the totals only of (*a*) the gross valuation; (*b*) retention; (*c*) amounts included in previous *certificates*; and (*d*) balance due for payment. Supporting information which should accompany the valuation recommendation are details of payments included in respect of nominated sub-contractors' work and full details of the retention deductions made. A standard form for this purpose also is published by the Royal Institution of Chartered Surveyors. Examples of its use are given in Figures 12.3, 12.6 and 12.9. The detailed build-up of valuations is not normally sent to the architect.

Liquidated and ascertained damages

It should be understood that no deduction should be made from the amount of a valuation for liquidated and ascertained damages (see p. 219) because of failure on the contractor's part to complete the Works on time. Even though the surveyor knows that the Employer is entitled to such damages, there is no provision in the

JCT Conditions of Contract for anyone other than the Employer himself to make the deduction from any sum due to the contractor.

The Employer may, if he so wishes, waive his right to damages, in which case he will make no deduction. It may, however, be prudent for the surveyor to remind the architect that damages are due to the Employer – and even how much – so that the architect may be in a position to advise the Employer accordingly.

Examples of interim valuations

The following examples illustrate many of the details which have been discussed earlier in the chapter. They relate to a contract of which particulars are given in Appendix A on pp. 336–341.

Valuation No. 1

This was done on 24 April 1994, approximately one month after the contractor took possession of the site (see Figures 12.1, 12.2 and 12.3). From an initial inspection of the Works, the surveyor found that the following work had been done:

About two-thirds of the demolition work and about 10 per cent of the excavation for the main building; some concrete bases were cast (items 27A–27C in Bill No. 3); about 10 per cent of the drainage work; water and electricity mains connections.

Mess huts, drying sheds, site offices and about half of the temporary fencing erected and temporary water supplies provided; about half of the temporary roads were laid.

Quantities of sand, aggregate and cement stored on site were estimated and noted down.

Commentary on Valuation No. 1

1   As part of the Preliminaries is cost-related, they cannot be calculated until the cost of the builder's work is known. This is indicated by inserting the words 'see below' opposite 'Bill No. 1 – Preliminaries' (Figure 12.1). See p. 183 for breakdown of Preliminaries.

SHOPS AND FLATS, THAMES ST., SKINTON

Valuation No. 1–24 April 1994	£ p	£ p
Bill No. 1 – Preliminaries		*See below*
Bill No. 2 – Demolition		3 531 00
Bill No. 3 – Shops and flats		
Groundwork say 10% of £31 256	3 126 00	
In situ concrete 27A–27C	3 267 00	
		6 393 00
Bill No. 4 – External works		
Drainage say 10% of £16 004		1 600 00
		11 524 00
Addition to correct for errors in bills of quantities	2.79%	321 00
		11 845 00
Preliminaries*		
Time-related items	1 337 00	
Fixed charges	2 088 00	
Cost-related items 2.5% of £11 845	296 00	3 721 00
		15 566 00
Statutory Bodies		
Thames Water – water main		
connection	350 00	
Seeboard – electric main		
connection	350 00	
	700 00	
profit 5%	35 00	
		735 00
Materials on site		
sand 10 tonnes £ 8.14	81 40	
aggregate 20 tonnes £ 7.92	158 40	
cement 9 tonnes £68.05	612 45	
		852 00
		17 153 00
Retention	17 153 00	
less amount subject to nil retention:		
Statutory Bodies	700 00	
	3% of 16 453 00	493 00
Total of Valuation No. 1		16 660 00

* *see* p. 183.

Figure 12.1

Valuation

Chartered Quantity Surveyor
Kewess & Partners
52 High Street, Urbiston

Works   Shops and Flats
28-34 Thames Street
Skinton

Valuation No: 1
Date of issue: 24th April 1994
QS Reference: 265/6

To Architect/Contract Administrator

Draw & Partners
25 Bridge Street
Skinton

Employer

Skinton Development Co.
High Path
Skinton

Contractor

Beecon Ltd
River Road
Skinton

Contract sum   £ 495,000.00

I/We have made, under the terms of the Contract, an Interim Valuation

as at   24th April 1994   * and I/we report as follows:—

Gross Valuation
(excluding any work or material notified to me/us by the Architect/The Contract Administrator in
writing, as not being in accordance with the Contract).

£   17153.00

Less total amount of Retention, as attached Statement.

£   493.00

£   16660.00

Less total amount stated as due in Interim Certificates previously issued by the
Architect/The Contract Administrator up to and including Interim Certificate No. ....

£   —

Balance (in words)   Sixteen thousand six hundred and
sixty pounds

£   16660.00

Signature: *C. Kewess.*

Chartered Quantity Surveyor FRICS/ARICS
(delete as applicable)

Notes:
(i)   All the above amounts are exclusive of V.A.T.
(ii)   The balance stated is subject to any statutory deductions which the Employer may be obliged to make under the provisions of
       the Finance (No. 2) Act 1975 where the Employer is classed as a 'Contractor' for the purposes of the Act.
(iii) It is assumed that the Architect/The Contract Administrator will:—
       (a) satisfy himself that there is no further work or material which is not in accordance with the Contract.
       (b) notify Nominated Sub-Contractors of payments directed for them and of Retention held therein by the Employer.
       (c) satisfy himself that the previous payments directed for Nominated Sub-Contractors have been discharged.
* (iv) The Architect's/The Contract Administrator's Interim Certificate should be issued within seven days of the date indicated thus
  (v)   Action by the Contractor should be taken on the basis of figures in, or attached to, the Architect's/The Contract
        Administrator's Interim Certificate.

© 1980 RICS

Figure 12.2

Statement of Retention and of Nominated Sub-Contractors' Values

Chartered Quantity Surveyor

Kewess & Partners
52 High Street, Urbiston

Works

Shops and Flats
28-34 Thames Street
Skinton

This Statement relates to:

Valuation No : 1
Date of issue : 24th April 1994
QS Reference : 265/6

	Gross Valuation	Amount subject to:			Amount of Retention	Net Valuation	Amount Previously Certified	Balance
		Full Retention of 3 %	Half Retention of 1½%	No Retention				
	£	£	£	£	£	£	£	£
Main Contractor	17 153	16 453	–	700	493	16 660	–	16 660
Nominated Sub-Contractors:-								
TOTAL	17 153	16 453	–	700	493	16 660	–	16 660

© 1980 RICS

No account has been taken of any discounts for cash to which the Contractor may be entitled if discharging the balance within 17 days of the issue of the Architect's/Contract Administrator's Interim Certificate.
The sums stated are exclusive of V.A.T.

Figure 12.3

2   For calculation of the percentage adjustment for errors in the bills of quantities, see p. 66. This item is inserted before the 'Preliminaries' are added because they were excluded when calculating the percentage.

Valuation No. 5

This was carried out on 25 August 1994, by which time five months of the contract period had passed (see Figures 12.4, 12.5 and 12.6). This time the surveyor noted that the following work had been done since the job began:

The piling sub-contractors had finished their work and had left the site; excavation was about 80 per cent complete; varying amounts of concrete work and brickwork had been done; woodwork in the roof of the single-storey part was being fixed; some windows, external doors and internal door frames were in position; plumbing and electrical work was in progress; about 45 per cent of the drainage and about 25 per cent of the permanent fencing had been done.

Quantities of sand, aggregate, cement, bricks, partition blocks, timber, felt damp proof course and wall ties were stored on site – these were estimated and noted down.

The architect had on 18 August, under clause 35.16, issued a certificate of Practical Completion of the piling sub-contract.

During the visit, the contractor's surveyor agreed the revised total of £1213 for increased costs of labour and materials, which included the 10 per cent addition under clause 39.8.* Three variation orders had been issued so far but it was agreed that not enough had been done on them to be taken into account in this valuation. The total of Certificates 1–4 inclusive was £99 842.00.

Commentary on Valuation No. 5

1   As part of the Preliminaries is cost-related, they cannot be calculated until the cost of the builder's work is known. This is indicated by inserting the words 'see below' opposite 'Bill No. 1 – Preliminaries' (Figure 12.4). See p. 183 for breakdown of Preliminaries.

---

* *see* p. 190 and Appendix B, p. 348

SHOPS AND FLATS, THAMES ST., SKINTON

Valuation No. 5 – 25 August 1994			£ p	£ p
Bill No. 1 – Preliminaries				See below
Bill No. 2 – Demolition				5 531 00
Bill No. 3 – Shops and Flats				
Groundwork	say 80% of £31 256		25 004 00	
In situ concrete	p. 27	7281		
	28A–28C	3819		
	29F–29H	3265		
	30A–30D	2026		
		———	16 391 00	
Masonry	35A–36F	9245		
	37C–37E	1573		
	42A–42D	2181		
		———	12 999 00	
Structural/Timber	47E–47H	833		
	52A–53F	978		
		———	1 811 00	
Waterproofing	80A–81C	406		
	89C–90A	701		
		———	1 107 00	
Winds/Doors/Stairs	93A–94D	1926		
	98B–98C	2105		
	99D–99G	1410		
		———	5 441 00	
Surface finishes	105A–106C	3550		
	108D–108E	270		
		———	3 820 00	
Piped supply system	125A–126C	6167		
	127C–127H	662		
	128A–128F	1635		
		———	8 464 00	
Electrical services	say 30% of £9389		2 817 00	
			———	77 854 00
Bill No. 4 – External works				
Drainage	say 45% of £16 004		7 202 00	
Fencing	say 25% of £17 244		4 311 00	
			———	11 513 00
				94 898 00
Addition to correct for errors in b/q		2.79%		2 647 00
Subtotal				97 545 00
Preliminaries				
Time-related items	5 × £1337		6 685 00	
Fixed charges			4 177 00	
Cost-related items	2.5% of £97 545		2 438 00	
			———	13 300 00
		Carried forward		110 845 00

Figure 12.4

			£ p	£ p
Brought forward				110 845 00
Nominated sub-contractors				
Drivas & Co. Ltd – piling			11 340 00	
Sparkes & Co. Ltd – electrical			5 000 00	
			16 340 00	
Profit	5%		817 00	
Attendances – piling			720 00	
– electrical			250 00	
				18 127 00
Statutory Bodies				
Thames Water – water main connection			350 00	
Seeboard – electric main connection			350 00	
Segas – gas main connection			500 00	
			1 200 00	
Profit	5%		60 00	
				1 260 00
Materials on site				
sand	20 tonnes	£ 8.14	162 80	
aggregate	30 tonnes	£ 7.92	237 60	
cement	4 tonnes	£ 68.05	272 20	
flettons	16m	£110.00	1 760 00	
partition blocks	90m²	£ 5.60	504 00	
damp proof course	20m	£ 0.86	17 20	
wall ties	400 no.	£ 4.65	18 60	
softwood	200 m	£ 0.69	138 00	
			3 110 00	
Fluctuations as agreed*			1 213 00	
				134 555 00
Retention			134 555 00	
less amounts subject to nil retention:				
Statutory Bodies	1 200			
Fluctuations	1 213			
			2 413 00	
			132 142 00	
less amount subject to half retention:				
Piling sub-contract			11 340 00	
			120 802 00	
3% of £120 802 00			3 624 00	
1½% of £11 340 00			170 00	
				3 794 00
Total of Valuation No. 5				130 761 00
Less total of Certificates Nos. 1–4				99 842 00
Total amount due				30 919 00

* including 10% addition as clause 39.8

Figure 12.4 continued

Valuation

Chartered Quantity Surveyor

Kewess & Partners
52 High Street, Urbiston

Valuation No:	5
Date of Issue:	25th August 1994
QS Reference:	265/6

Works Shops and Flats
28-34 Thames Street
Skinton

To Architect/Contract Administrator

Draw & Partners
25 Bridge Street
Skinton

Employer

Skinton Development Co.
High Path
Skinton

Contractor

Beecon Ltd
River Road
Skinton

Contract sum £ 495,000

I/We have made, under the terms of the Contract, an Interim Valuation

as at 25th August 1994 * and I/we report as follows:—

Gross Valuation
(excluding any work or material notified to me/us by the Architect/The Contract Administrator in writing, as not being in accordance with the Contract).

£ 134 555.00

Less total amount of Retention, as attached Statement.

£ 3 794.00

£ 130 761.00

Less total amount stated as due in Interim Certificates previously issued by the
Architect/The Contract Administrator up to and including Interim Certificate No. ...4...

£ 99 842.00

Balance (in words) Thirty thousand nine hundred and nineteen pounds

£ 30 919.00

Signature: *C. Kewess*

Chartered Quantity Surveyor FRICS/ARICS
(delete as applicable)

Notes:

(i) All the above amounts are exclusive of V.A.T.
(ii) The balance stated is subject to any statutory deductions which the Employer may be obliged to make under the provisions of the Finance (No. 2) Act 1975 where the Employer is classed as a 'Contractor' for the purposes of the Act.
(iii) It is assumed that the Architect/The Contract Administrator will:—
(a) satisfy himself that there is no further work or material which is not in accordance with the Contract.
(b) notify Nominated Sub-Contractors of payments directed for them and of Retention held therein by the Employer.
(c) satisfy himself that the previous payments directed for Nominated Sub-Contractors have been discharged.
* (iv) The Architect's/The Contract Administrator's Interim Certificate should be issued within seven days of the date indicated thus
M Action by the Contractor should be taken on the basis of figures in, or attached to, the Architect's/The Contract Administrator's Interim Certificate.

© 1990 RICS

Figure 12.5

Statement of Retention and of Nominated Sub-Contractors' Values

Chartered Quantity Surveyor
Kewess & Partners
52 High Street, Urbiston

Works   Shops and Flats
28–34 Thames Street
Skinton

This Statement relates to:
Valuation No : 5
Date of issue : 25th August 1994
OS Reference : 265/6

	Gross Valuation	Amount subject to:			Amount of Retention	Net Valuation	Amount Previously Certified	Balance
		Full Retention of 3%	Half Retention of 1½%	No Retention				
	£	£	£	£	£	£	£	£
Main Contractor	118 215	115 802	–	2413	3474	114 741	89 707	25 034
Nominated Sub-Contractors:-								
1. Drivas & Co. Ltd - piling	11 340	–	11 340	–	170	11 170	10 135	1 035
2. Sparkes & Co. Ltd - elec.	5 000	5 000	–	–	150	4 850	–	4 850
TOTAL	134 555	120 802	11 340	2413	3794	130 761	99 842	30 919

No account has been taken of any discounts for cash to which the Contractor may be entitled if discharging the balance within 17 days of the issue of the Architect's/Contract Administrator's Interim Certificate. The sums stated are exclusive of V.A.T.

Figure 12.6

2 For calculation of the percentage adjustment for errors in the bills of quantities, see p. 66. This item is inserted before the 'Preliminaries' are added because they were excluded when calculating the percentage.
3 The fixed charges part of Preliminaries is the total of 'Fixed charges (a)' on p. 183.
4 Attendances on the electrical sub-contractor are in proportion to the value of work done on the sub-contract.
5 The amounts subject to half or nil retention are, in the case of nominated sub-contractors and statutory bodies, exclusive of profit and attendance, as those amounts are paid to the main contractor.
6 The deduction for previous payments should be the total of the architect's certificates, not of the surveyor's valuations, as it is possible for them to differ.

Valuation No. 12

This was done on 25 March 1995, that is, after twelve months had elapsed (see Figures 12.7, 12.8, and 12.9). Due to bad weather, delays in deliveries because of strikes in the haulage industry during the early part of the year and the delay in installing the lifts, the contract was about five weeks behind schedule.

The five flats on the second floor had been recently completed, those on the first floor and the staircase were in the early stages of plastering, whilst the shops on the ground floor were at first fixings stage. Drainage was finished and the fencing was about two-thirds complete. About ten per cent of the external pavings had been laid.

Increased costs of labour and materials had been agreed within the past week at £2364, including ten per cent addition under clause 39.8. Thirteen variations had been issued to date, one since the last valuation.

On 21 March, by agreement with the contractor, the Employer had taken possession of the five flats on the second floor, although the lifts were not yet in operation. On 22 March, the architect issued a written statement identifying the part taken over and giving the date of possession, as required by clause 18.1 of the JCT Form. The surveyor estimated the total value of the five flats to be £80,000 for release of retention purposes.

SHOPS AND FLATS, THAMES ST., SKINTON

Valuation No. 12–25 March 1995			£ p	£ p
Bill No. 1 – Preliminaries				*See below*
Bill No. 2 – Demolition				5 531 00
Bill No. 3 – Shops and Flats				
Groundwork	say 90% of	£31 256	28 130 00	
In situ concrete	pp. 27–30	25 675		
	31A–31E	9 171		
	32C	1 476		
			36 322 00	
Masonry	pp. 34–40	26 743		
	42A–42F	4 716		
	43A	295		
			31 754 00	
Structural/Timber	47A–47H	4 251		
	50A–52J	3 726		
	53C–58F	4 273		
			12 250 00	
Cladding/Covering	70A–73C	3 906		
	78A–79D	1 374		
			5 280 00	
Waterproofing	80A–86C	3 421		
	89C–91B	2 891		
			6 312 00	
Winds/Doors/Stairs	93A–96C	13 274		
	98B–98D	1 485		
	99D–100B	3 539		
			18 298 00	
Surface finishes	say 55% of	£15 100	8 305 00	
Bldg. fabric sunds.	113A–114D	1 372		
	115A–116A	4 510	5 882 00	
Disposal systems	say 60% of	£ 8 326	4 995 00	
Piped supply system	say 40% of	£19 149	7 657 00	
				165 185 00
Bill No. 4 – External works				
Drainage	138A–140F		16 004 00	
Fencing	say 60% of	£17 244	10 346 00	
Pavings	say 10% of 142A–142K		804 00	
				27 154 00
				197 870 00
Addition to correct for errors in b/q			2.79%	5 520 00
			Carried forward	203 390 00

Figure 12.7

Valuation No. 12 (cont.)

		£ p	£ p
	Brought forward		203 390 00
Preliminaries*			
Time-related items	12 × £1273†	15 276 00	
Fixed sums (a)	4 177 00		
(b)	248 00		
Cost-related items	2.5% of £203 390	4 425 00	
		5 084 00	
			24 785 00
Variations	as Valuation No. 11	3 373 00	
	AI No. 62	228 00	3 601 00
Nominated sub-contractors			
Drivas & Co. Ltd – piling		11 340 00	
Sparkes & Co. Ltd – electrical		10 700 00	
Asphaltank Ltd – asphalt work		12 000 00	
Wayward Lifts Ltd – lifts		6 300 00	
Smith Alarms Ltd – security system		1 250 00	
		41 590 00	
Profit	5%	2 079 00	
Attendances – piling	720		
– electrical	175		
– asphalt work	393		
– lifts	200		
– security system	–		
		1 488 00	
			45 157 00
Nominated suppliers:			
Kitchen fitments		1 000 00	
Ironmongery		1 350 00	
Sanitary appliances		2 000 00	
		4 350 00	
Profit	5%	217 00	
			4 567 00
Statutory bodies:			
Thames Water – water main connection		371 00	
Seeboard – electric main connection		380 00	
Segas – gas main connection		549 00	
Urbiston Council – sewer connections		922 00	
British Telecom – telephone		326 00	
		2 548 00	
Profit	5%	127 00	2 675 00
	Carried forward		284 175 00

* *see* p. 183.
† reduced from £1337 to allow for one month's delay

Figure 12.7 continued

Valuation No. 12 (cont.)

			£ p	£ p
		Brought forward		284 175 00

Materials on site:

			£ p	
cement	8 tonnes	£ 68.05	544 00	
fletton bricks	2 m	£110.00	220 00	
partition blocks	60 m²	£ 5.60	336 00	
doors	50 No.	£ 12.34	617 00	
plumbing goods	say		400 00	

	£ p	£ p
		2 117 00
Fluctuations, as agreed*		2 364 00
Claim under clause 26, as agreed		468 00
		289 124 00

	£ p	£ p
Retention	289 124 00	

less amounts subject to nil retention:

		£ p	£ p
Statutory Bodies	2 548		
Piling Sub-cont.	11 340		
Fluctuations	2 364		
Claim	468		
		16 720 00	
		272 404 00	

less amounts subject to half retention:

		£ p	£ p
Flats handed over 21/3/95:	80 000		
less proportion of piling s/c	1 832	78 168 00	
		194 236 00	
3% of £194 236 00		5 827 00	
1½% of £78 168 00		1 172 00	
			6 999 00

	£ p	£ p
Total of Valuation No. 12		282 125 00
Less total of Certificates Nos. 1–11		247 381 00
Total amount due		34 744 00

* including 10% addition as clause 39.8

Figure 12.7 continued

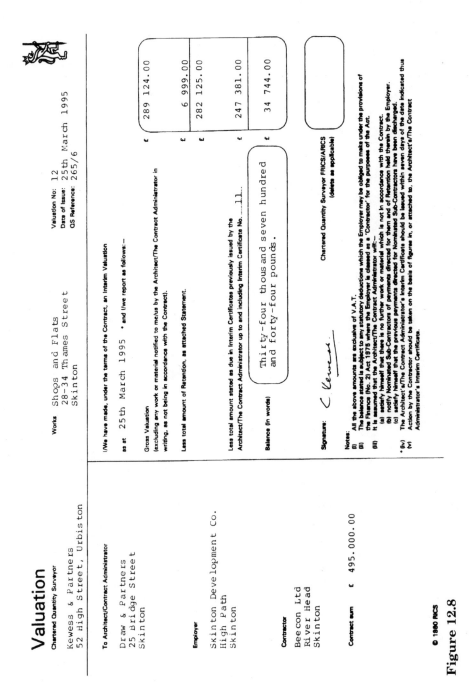

Valuation

Chartered Quantity Surveyor

Kewess & Partners
52 High Street, Urbiston

Works Shops and Flats
28-34 Thames Street
Skinton

Valuation No: 12
Date of Issue: 25th March 1995
QS Reference: 265/6

To Architect/Contract Administrator

Draw & Partners
25 Bridge Street
Skinton

I/We have made, under the terms of the Contract, an Interim Valuation

as at   25th March 1995   * and I/we report as follows:-

Gross Valuation
(excluding any work or material notified to me/us by the Architect/The Contract Administrator in writing, as not being in accordance with the Contract).

£   289 124.00

Less total amount of Retention, as attached Statement.

£   6 999.00

£   282 125.00

Less total amount stated as due in Interim Certificates previously issued by the
Architect/The Contract Administrator up to and including Interim Certificate No. ....11....

£   247 381.00

Balance (in words)

Thirty-four thousand and seven hundred and forty-four pounds.

£   34 744.00

Employer

Skinton Development Co.
High Path
Skinton

Contractor

Beecon Ltd
River Head
Skinton

Contract sum   £   495.000.00

Signature: C. Kewess

Chartered Quantity Surveyor FRICS/ARICS
(delete as applicable)

Notes:

(i) All the above amounts are exclusive of V.A.T.
(ii) The balance stated is subject to any statutory deductions which the Employer may be obliged to make under the provisions of the Finance (No. 2) Act 1975 where the Employer is classed as a "Contractor" for the purposes of the Act.
(iii) It is assumed that the Architect/The Contract Administrator will:—
(a) satisfy himself that there is no further work or material which is not in accordance with the Contract.
(b) notify Nominated Sub-Contractors of payments directed for them and of Retention held therein by the Employer.
(c) satisfy himself that the previous payments directed for Nominated Sub-Contractors have been discharged.
* (iv) The Architect's/The Contract Administrator's Interim Certificate should be issued within seven days of the date indicated thus
(v) Action by the Contractor should be taken on the basis of figures in, or attached to, the Architect's/The Contract Administrator's Interim Certificate.

© 1980 RICS

Figure 12.8

Statement of Retention and of Nominated Sub-Contractors' Values

Chartered Quantity Surveyor
Kewess & Partners
52 High Street, Urbiston

Works   Shops and Flats
28–34 Thames Street
Skinton

This Statement relates to:
Valuation No : 12
Date of issue : 25th March 1995
OS Reference: 265/6

	Gross Valuation	Amount subject to:			Amount of Retention	Net Valuation	Amount Previously Certified	Balance
		Full Retention of 3%	Half Retention of 1½%	No Retention				
	£	£	£	£	£	£	£	£
Main Contractor	247 534	163 986	78 168	5 380	6091	241 443	214 022	27 421
Nominated Sub-Contractors:								
1. Drivas & Co. Ltd - piling	11 340			11 340	–	11 340	11 340	–
2. Sparkes & Co. Ltd - electrical	10 700	10 700	–	–	321	10 379	6 887	3 492
3. Asphaltank Ltd - asphalt work	12 000	12 000	–	–	360	11 640	9 021	2 619
4. Wayward Lifts - lifts	6 300	6 300	–	–	189	6 111	6 111	–
5. Smith Alarms Ltd - security system	1 250	1 250	–	–	38	1 212	–	1 212
TOTAL	289 124	194 236	78 168	16 720	6999	282 125	247 381	34 744

No account has been taken of any discounts for cash to which the Contractor may be entitled if discharging the balance within 17 days of the issue of the Architect's/Contract Administrator's Interim Certificate. The sums stated are exclusive of V.A.T.

© 1980 RICS

Figure 12.9

In January, the architect had agreed to a request to release the retention on the piling sub-contract, and this had been effected in Certificate No. 11.

Just before the February valuation date, it came to light, as a result of a complaint from the lifts sub-contractor, that the main contractor had failed to make payment of the sum of £6111 due to the sub-contractor and included in Certificate No. 10. In consequence, work on the lifts installation had stopped and the money had still not been paid. The architect had issued a certificate under clause 35.13.5.2 and the Employer had subsequently made the payment direct, under the provision of that clause.

During February a claim had been made by the contractor for loss arising under clause 26, due to the architect's failure to provide a schedule of the decorations and details of selected wallpapers required in the flats, thus delaying the completion of the second floor flats. The claim, amounting to £468, was accepted and agreed on 15 March.

The total of Certificates Nos. 1–11 was £247,381.00.

Commentary on Valuation No. 12

1 As part of the Preliminaries is cost-related, they cannot be calculated until the cost of the builder's work is known. This is indicated by inserting the words 'see below' opposite 'Bill No. 1 – Preliminaries' (Figure 12.7). See p. 183 for breakdown of Preliminaries.

2 For calculation of the percentage adjustment for errors in the bills of quantities, see p. 66. This item is inserted before the 'Preliminaries' are added because they were excluded when calculating the percentage.

3 The monthly amount of the time-related portion of the Preliminaries has been recalculated to allow for one month's delay.

4 The fixed charges part of the Preliminaries is the total of 'fixed charges (a)' and 25 per cent of the amounts for 'drying out' and 'cleaning' in col. (b) in respect of the part handed over on 21 March 1995 (see p. 183).

5 Attendances on nominated sub-contractors are in proportion to the values of work done.

6 The amounts for statutory bodies are the rounded-down sums shown on pp. 95–6, except that, by the valuation date, only the first four accounts for Employer's telephone call charges had been received from British Telecom.

7 The amounts subject to half or nil retention are, in the case of nominated sub-contractors and statutory bodies, exclusive of profit and attendance, as those amounts are paid to the main contractor.

8 The total value of the flats handed over on 21 March 1995 estimated at £80 000 must be reduced by a proportion of the value of the piling sub-contract, for which retention has already been released. In later valuations, the total will have to be further reduced by a similar proportion of other nominated sub-contracts released from retention. It is probably best therefore to reduce the total value of the 'relevant part' by the appropriate proportion of the total of all the nominated sub-contracts (except any relating to external works) in all valuations subsequent to the date of handing over. The proportion is:

$$\frac{\text{gross value of the 'relevant part' } \times \text{ value of the sub-contract}}{\text{total contract value}}$$

In this case the calculation of the reduction in relation to the piling sub-contract only is

$$\frac{80\ 000 \times 11\ 340}{495\ 000} = £1832.$$

9 The deduction for previous payments should be the total of the architect's certificates, not of the surveyor's valuations, as it is possible for them to differ.

10 No deduction should be made of the amount paid direct by the Employer to Wayward Lifts (see p. 209), nor for any liquidated damages which may be due because of the five weeks' delay. However, the surveyor should draw the architect's attention to such deductions so that he may remind the Employer of his right to deduct from the payment stated in the certificate as due.

References

1. JCT, *Standard Form of Building Contract, 1980 Edition*, clause 30.2.
2. TURNER, Dennis F., *Quantity Surveying Practice and Administration* (London: George Godwin Ltd, 1983), p. 159.
3. JCT, *Standard Form of Building Contract, 1980 Edition*, Articles of Agreement, Second Recital.
4. The Query Sheet, *Chartered Surveyor, BQS Quarterly*, Vol. 1, No. 2, December 1973, p. 40.
5. TURNER, Dennis F., *Quantity Surveying Practice and Administration* (London: George Godwin Ltd, 1983), p. 164.
6. GOODACRE, P. E., *Formula Method of Price Adjustment for Building Contracts* (Reading: College of Estate Management, 1978), p. 96, Q. 2.
7. JCT, *Standard Form of Building Contract, 1980 Edition*, clause 36.4.4.
8. ibid., clause 36.3.1.

13

Final accounts

In order to be able to make final payment to a contractor of the sum to which he is entitled under the terms of the contract, it is usually necessary to produce a final account. In any but the very smallest jobs, there will be many adjustments to be made to the Contract Sum and a detailed document is necessary to show, to the satisfaction of all concerned, what amount of money the Employer is liable to pay and the contractor is entitled to receive and how it has been calculated.

In the case of lump sum contracts, therefore, the final account begins with the Contract Sum, shows what amounts are deducted and what amounts are added (and for what reasons) and ends with the adjusted total sum. This sum, when agreed by the contractor, is the total amount which the Employer will pay to the contractor for the work he has done.

In the case of measured contracts, the final account is built up from nil to an 'ascertained final sum', which is the aggregate of amounts for named parts of the project. The ascertained final sum is the total amount which the Employer will pay to the contractor.

The responsibility for preparing the final account is that of the quantity surveyor and the JCT Form limits the time in which the task should be completed. As previously stated (see p. 110), the surveyor has a duty to endeavour to complete the variation accounts within the period stated in clause 30.6.1.2, and the whole account in time for the architect to issue the Final Certificate at the time specified in clause 30.8.

It will be necessary for the surveyor to obtain the contractor's agreement to the total of the account. To be able to give this, the contractor will need to see the supporting details. Accordingly, a photocopy of the whole document, fully priced and totalled, should be sent to the contractor, either complete or in sections as they are finished. As there will often be some points of disagreement, the surveyor and the contractor's surveyor will need to meet and discuss those points, negotiate revisions to any suggested item rates and prices, etc., until eventually the whole is agreed. Each of them (or their principals) will then sign the summary page to signify their agreement.

Constituents of final accounts

Lump sum contracts (except those based on bills of approximate quantities)

The final account will usually consist of a document which contains most or all of the following sections, although not necessarily in the order given:

(*a*) a summary of the account;
(*b*) adjustments of prime cost sums;
(*c*) adjustments of provisional sums;
(*d*) variation accounts;
(*e*) adjustment of provisional items;
(*f*) claims;
(*g*) fluctuations in costs of labour, materials and statutory contributions, levies, and taxes.

The document may be handwritten or it may be typed, depending on the client's requirements. As multiple copies are not usually needed, the expense of typing seems unjustified. The client is often only concerned with the summary and, if that is so, that is all that needs to be typed. A photocopy of the handwritten document will provide a copy for the contractor's perusal and agreement as suggested above.

The foregoing list of sections of the account require comment as follows.

Summary of the account

The summary usually consists of a single page of double billing paper showing the contract sum, the respective total deduction and addition amounts transferred from each of the succeeding sections, and the final total. At the bottom of the page will appear the signatures of the contractor and surveyor. A suitable format is shown in Figure 13.1.

Adjustments of prime cost sums

The summary of the adjustments forms the first page of the section, the following pages showing complete details of the deductions and additions respectively of the p.c. sums and the actual costs, with the profit amounts. Figures 7.1 and 7.2 show alternative formats for the summary. Figure 7.1 illustrates a method of setting out the supporting details.

Adjustments of provisional sums

This section will be similar in layout to the preceding one, except that there will be no separate profit amounts. The additions items set against each provisional sum will normally be measured and priced in exactly the same way as those in variation accounts, where the work has been carried out by the general contractor as in the majority of cases.

Variation accounts

The variation accounts (see p. 110) may commence with a summary showing the total of deductions and the total of additions for each separate variation or, alternatively, the net total saving or extra for each. Following the summary, the supporting details are given consisting of measured items grouped as omissions and additions under each AI. A third format shows only the total net saving or extra for all the variations together, and the presentation of the details is in the form of a bill of omissions and a bill of additions without items being separated under each variation.

Adjustment of provisional items

This section of final accounts requires more detailed consideration as, unlike the other sections, it does not form a separate chapter.

When the quantity of an item or group of items of work cannot be accurately ascertained at the time of preparing the tender documents, the item or group of items is distinguished by being described as 'provisional'. In the case of single items, it will appear in brackets at the end of the description. In the case of a group of items, for example, those for the substructure of a building, the words 'The following "All Provisional"' will normally precede the group.

In all cases, whether single items or groups of items, it is incumbent upon the surveyor to deduct the value of the provisional items from the contract sum and to substitute the value of measured items as actually carried out. This procedure is required by clauses 30.6.2.2 and 30.6.2.12 of the JCT Form. Even where the quantity of an item as executed is the same as that in the bills, the deduction and addition should still be shown. Otherwise, it may be supposed that the provisional item has been overlooked. The value of the remeasured items will be ascertained in the same way as in the case of variations (see pp. 111–3).

The advantage of giving measured items in the bills even though the quantities are not accurate, rather than allowing a provisional sum instead, is first, that unit rates are then available for use in valuing the work as remeasured. Secondly, the contractor is given a much more definite picture of the scope and nature of the work than he would otherwise have.

Claims

Claims for additional payments which have been accepted under clauses 26.1 and 34.3 of the JCT Form should each be fully and separately set out, detailing the calculations of the agreed sums. If the section is large enough to warrant it, a summary of the claims should be presented on its first page.

Fluctuations in costs of labour and materials, etc

This is a particular form of claim arising under clauses 38, 39 and 40 of the JCT Form. The details should be fully set out to show

how the amount of the claim has been calculated (see Chapters 9 and 10).

Measured contracts (that is, those based upon bills of approximate quantities or schedules of rates)

The final account for contracts which require the complete measurement of the Works will usually be a simpler, though more lengthy document. As there is no contract sum, the account will not be an adjustment account (i.e. showing deductions and additions), but will consist wholly of additions amounts.

Clause 30.6.2 of the JCT Form for use with bills of approximate quantities lists the components of the 'Ascertained Final Sum' and they may be conveniently grouped together as follows. The same groupings may also be applied to schedule of rates contracts, as appropriate.

Summary of the account

The summary cannot commence with a contract sum as there is none, nor in consequence can it show any omissions. Instead, it will give a list of the succeeding sections with a total sum against each.

Bills of remeasurement

These will be similar in format to the measured sections of bills of quantities prepared for tendering purposes.

Preliminaries

This section will consist of the appropriate preliminaries items, valued in accordance with the Conditions of the contract.

Nominated sub-contractors', nominated suppliers', and statutory bodies' accounts

This section will be similar to the 'adjustment of p.c. sums' section of the account for a lump sum contract, but without the omissions amounts.

Claims

This section will be similar to that in an account for a lump sum contract (see under 'Claims', p. 215).

Fluctuations

This section will also be similar to that in an account for a lump sum contract (see under 'Fluctuations in costs of labour and materials', p. 215).

Prime cost contracts

The final account for prime cost contracts is in essence a very simple document. It shows the total of each of the constituents of the prime cost, i.e. labour, plant, and materials, and, in addition, either the fixed fee or the calculation of the fee where it is a percentage of the prime cost. If there is a target cost (see pp. 20–1), the resulting addition or deduction should also be shown. In the event of a situation arising being such as to warrant an adjustment of the fee,* details of the calculation of the adjustment should also be given.

Final certificate

When the final account has been agreed, a photocopy of the original document (or, if sufficient, just the summary pages) will be sent to the architect or the Employer (if appropriate) with a notification of the balance due to the contractor in settlement, after deduction of the total amount of interim certificates. It may be necessary to prepare a separate statement to show the reduced sum due, if the Employer has made payments when the contractor has defaulted. Examples are: where the contractor has failed to pay insurance premiums (clauses 21 and 22A of the JCT Form) and where the contractor has failed to pay nominated sub-contractors the sums included in interim certificates (clause 35.13). (It should be noted that the Form does not require such

---

* *See* JCT Standard Form of Prime Cost Contract.

amounts to be deducted from the contract sum, i.e. in the final account, but to be deducted by the Employer from 'any sums due or to become due to the contractor'.)

Upon notification of the sum due, the architect is required to issue the Final Certificate within two months after the happening of the last of the events listed in clause 30.8. This should state (*a*) the adjusted contract sum or the ascertained final sum (i.e. the total of the final account), (*b*) the total of amounts paid on account and (*c*) the difference between the two amounts expressed as a

SHOPS AND FLATS, THAMES ST., SKINTON
for
SKINTON DEVELOPMENT CO.
FINAL ACCOUNT

	Omissions	Additions
	£ p	£ p
Contract sum as contract dated 23rd February 1994		495 000 00
Adjustment of prime cost and provisional sums	8 486 34	
Variations		6 167 30
Remeasurement of provisional work	849 09	
Fluctuations in costs of labour and materials		5 157 61
Claim for loss under clause 26		468 00
	9 335 43	506 792 91
Less omissions		9 335 43
TOTAL OF FINAL ACCOUNT		497 457 48

Signed for and on behalf of	Signed for and on behalf of
Beecon Ltd,	Kewess & Partners
River Road,	52, High Street,
Skinton,	Urbiston,
Middlesex.	Surrey.

. . . .

15 June 1996

Figure 13.1 Final account summary

balance due from one party to the other. Such balance is stated to be 'subject to any deductions authorized by the Conditions', such as those instanced in the preceding paragraph.

Liquidated and ascertained damages

Clause 24.2 of the JCT Form provides for liquidated damages to be paid or allowed by the contractor to the Employer calculated at the rate stated in the Appendix to the Conditions, when a contract exceeds the contract period or, if an extension has been granted, the extended contract period.

Clause 24.2 says that the Employer may deduct any sum for liquidated damages from 'any monies due or to become due to the contractor'. As far as the surveyor is concerned, therefore, the deduction of any amount for liquidated damages due should be made by the Employer from the sum stated in a certificate issued by the architect, as due to the contractor. All that the surveyor should do is to notify the architect of the amount of liquidated damages payable, so that the architect may in turn inform the Employer of the amount which he is entitled to deduct, if he so wishes. The surveyor should not deduct liquidated damages in the final account.

Example of final account

Figure 13.1 shows a method of setting out the summary of the final account for the contract detailed in Appendix A. The supporting details relate to the examples given in Chapters 7–11.

14

Cost control

The surveyor's role

Cost control has been defined as 'the controlling measures necessary to ensure that the authorized cost of the project is not exceeded'.[1] Initially, the 'authorized cost' is the contract sum, but at subsequent times in the construction period it will need to be adjusted to take account of necessary savings and of additional costs which the Employer is willing and able to meet.

The 'controlling measures' are usually in the hands of the Employer and his architect. The surveyor is not normally in a position to exercise control of cost himself unless, as is sometimes the case, he has been appointed project manager, when he has overall control of the project. His usual role, however, is that of adviser to those who have that control and so is one of monitoring and reporting on costs rather than controlling them.

The contractor's surveyor likewise is unlikely to have any direct control of the costs of construction but is in the position of monitoring costs and expenditure and advising the contracts manager on such matters as the value of work done to date and the likely eventual total value of the contract. The contracts manager will be concerned with the relationship between the contractor's expenditure on the construction work and the total of payments received or anticipated from the Employer. While the payments are in excess of the expenditure, there is a profit. If in any part of the work, however, the contractor's costs exceed the amounts payable under the contract terms, there will exist a loss situation, at least so far as that section of the work is concerned. The contractor's surveyor will be expected to be able to explain why this situation has arisen and may even be expected to foresee a potential loss-making position developing and to warn the contracts manager so that he may take what steps are necessary to minimize or reverse the loss.

Such a position may arise from any one or more of the following causes:

(a) inefficient deployment of resources (labour, plant, and materials);
(b) excessive wastage or theft of materials;
(c) plant being allowed to stand idle or under-utilized;
(d) adverse weather or working conditions;
(e) delays arising from one or more of a variety of causes;
(f) under-pricing of tender documents by assumptions in regard to labour times, types and sizes of plant, etc., which do not equate with the realities of the construction work.

Recovery of additional expenditure under some of the foregoing heads may be possible (e.g. items (c), (d) and (e) on the list) by successful claims within the provisions of the contract but this will not be so in all cases.

Cost control from the viewpoint of the contractor's surveyor, therefore, is distinct from that of the surveyor. Although the concern of both of them centres around the total value of a project, the former is concerned with the relationship between the contractor's expenditure and payments to him by the Employer, while the latter is concerned with the relationship between the total amount that the Employer is willing and able to spend and the total sum that he is liable to pay under the terms of the contract.

In order to provide a basis for controlling cost, there must be a contract sum or an estimated value of the initial contract. In the case of lump sum contracts, there will be a contract sum, but in the case of prime cost and schedule of rates contracts there will not, and it will be necessary for the surveyor to prepare an estimate of contract value, preferably in the form of a cost plan. This will provide the basis for the necessary control mechanism, if cost is to be controlled at all.

Monitoring and reporting on the financial position

The Employer will need to be kept informed by the surveyor of the financial position of the contract. This is so that he will be able to make the necessary arrangements to have the finance available to pay the contractor within the stipulated time upon presentation of interim certificates and also so that he may be warned of any possibility or likelihood of the 'authorized cost' of the project being exceeded. He may then take steps, as considered appropriate, to increase his borrowing capacity or to effect savings on the project to ensure that the final cost is within his budget limit.

There are thus two aspects to financial reports: (*a*) the current position, and (*b*) the likely total cost. In order to carry out his duties adequately, the surveyor will have to take account of all the actual or likely causes of adjustment to the contract sum (as far as he is able to do so), as follows:

(*a*) expenditure of p.c. sums;
(*b*) expenditure of provisional sums;
(*c*) variations;
(*d*) claims;
(*e*) recoverable fluctuations in costs.

Expenditure of p.c. sums

The position at any time with regard to the expenditure of p.c. sums will usually be clear cut. The architect either will or will not have issued instructions to the contractor on each of the p.c. sums and, where he has, the amount of the saving or extra will be known. Where he has not taken a decision, then no adjustment of the p.c. sum can be made, unless it is obvious that it is going to be exceeded. In that case, the surveyor should include in his financial statement an estimated addition to that p.c. sum. The consequent effect (if any) on the amounts for profit and attendances in relation to each p.c. sum will also be known to the same extent.

Expenditure of provisional sums

Where provisional sums are to cover the cost of such matters as telephone calls made on behalf of the Employer or daywork, then estimates of the likely expenditure will have to be included if actual costs are not known. Where provisional sums are to cover work to be carried out by the general contractor, the actual cost of the work (if already carried out and measured and valued) should be substituted for the respective provisional sums. If the work has not yet been measured and valued, then an estimate of the likely cost should be made and substituted for the provisional sums. Such estimates will usually be done by the 'approximate quantities' method, i.e. by measuring the main items of work involved reasonably accurately (but not necessarily exactly). The measured items are then priced at 'all-in' rates, i.e. at rates (based on bill rates where possible) which allow not only for all the work described in each item but also for the cost of all related items which have not been separately measured.

Variations

Where variations have been included in Architect's Instructions and have been accurately measured and valued, the net extra cost or saving (i.e. taking both additions and omissions into account) should be incorporated in the financial assessment. In regard to those variations which have not been measured and valued, including any which are expected but have not yet been ordered by the architect, the total net value of each should be estimated in the manner described under 'Expenditure of provisional sums' and be taken into account.

Claims

If claims can be dealt with promptly, the amounts allowed can be taken into account when the next financial statement is made. Claims which have been presented but are still outstanding should be valued as accurately as reasonably possible (erring, if anything, on the generous side) and included in the statement. If any claim is subsequently rejected, it will, of course, be omitted from the next statement. Any claims unlikely to be accepted can probably be safely ignored.

Fluctuations

Fluctuations which are recoverable under the terms of the contract should be valued at least monthly for inclusion in interim valuations. Where the Formula Method is used, fluctuations will be calculated at each valuation date anyway (see p. 152).

The amounts so agreed will be included in financial reports and an estimate of future payments will need to be made for the purpose of forecasting the likely total cost of the project.

It may be thought that interim valuations are, in effect, reports on the current financial position of a project. While to a large extent this is so, they do not necessarily give a complete picture. They give the accurate current position of expenditure by the Employer, but they do not necessarily include all known costs for which he is currently liable. The position is also clouded to some extent by the retention, particularly where there have been some releases, and also by the inclusion in valuations of the value of unfixed materials. Examples are given below of expenditure to which the Employer is committed but which may not be included in the latest interim valuation, often because of time lag. The value of unfixed materials, included in interim valuations, would be excluded from financial reports.

(a) work covered by p.c. sums which has not yet been completed or of which the value has not yet been included in an interim valuation;

(b) work which is covered by provisional sums or which is the subject of variation orders, which is still in progress but was insufficiently advanced for inclusion in the last valuation;

(c) fluctuations for which the Employer is liable but for which supporting details are awaited from the contractor;

(d) claims which are expected to be accepted but which have not yet been settled.

The contractor's surveyor will be involved in preparing similar financial reports for the contracts manager, contracts director or managing director. His reports will, of course, be similar in most respects to those prepared by the Employer's surveyor. They may differ, however, in the following matters.

(*a*) Variations may be included which the contractor claims have been ordered but which have not been confirmed in writing. These may or may not be confirmed later (see pp. 106–7).

(*b*) The value of all claims will be included, whether accepted or not. A distinction may be shown, however, between those agreed or likely to be agreed and those unlikely to be accepted.

(*c*) The valuation of variations, remeasured work, claims and fluctuations will differ, partly because of item rates and prices which remain to be agreed and partly because aspects of claims and fluctuations payments may be the subject of dispute.

Examples of financial reports are given in Figures 14.1 and 14.2.

As the effective control of cost lies with the architect, it is desirable that the surveyor should bring his influence to bear to encourage the exercise of that control. He can do so by providing cost reports on the current and future financial positions which are prompt, reliable and up-to-date. It is also desirable that, before issuing instructions that affect the cost of the contract, the architect should consult the surveyor on the likely effect and obtain advice on the cost of any alternative that may be available.

Forecasting client's capital expenditure flow

If a contract is large, with a cost of millions of pounds to be expended over several years, it is essential to make a reasonably reliable forecast of the likely flow of expenditure. Even where costs are more modest, and contract time correspondingly shorter, it is to a client's advantage to be able to anticipate his cash flow requirement and to be able to arrange for finance to be available accordingly.

In consequence, it is often part of the surveyor's task to prepare, either in tabular or graphical form, an anticipated 'rate of spend' or 'expenditure flow' forecast.

In order to do this, it is necessary to prepare a programme for the carrying out of the various parts of the construction work.

SHOPS & FLATS, 28–34 THAMES ST., SKINTON

FINANCIAL REPORT NO. 10

Current position as at 2 April 1995

	£
Value of measured work as bills of quantities as at 25 March 1995	203 390
Value of Preliminaries as at 25 March 1995	24 785
Payments to nominated sub-contractors plus profit and attendances as Valuation No. 12	45 157
Payments to nominated suppliers plus profit as Valuation No. 12	4 567
Payments to Statutory Bodies plus profit as Valuation No. 12	2 675
Variations	3 601
Fluctuations in labour and materials costs	2 364
Claim under clause 26, as agreed	468
Claim under clause 26 outstanding but for which liable	2 500
VALUE OF WORK CARRIED OUT TO DATE	289 507
Total actual expenditure as Certificate No. 12 (before deduction of retention)	289 124
Total anticipated expenditure to date as forecast*	341 223

5 April 1995

* *see* p. 232.

Figure 14.1 Example of report on current financial position of a contract

SHOPS & FLATS, 28–34 THAMES ST., SKINTON

FINANCIAL REPORT NO. 11

Estimated Total Cost

	£
Contract sum	495 000
Less Contingency sum	8 000
	487 000
Less Prime cost and provisional sums (net omission)	4 536
	482 464
Variations	6 000
Fluctuations in labour and materials costs	5 000
Claims	2 968
	496 432
Contingency sum for remainder of contract	3 500
ESTIMATED TOTAL COST OF CONTRACT	499 932

5 April 1995

Figure 14.2 Example of report on likely total cost of a contract

A progress chart or 'bar chart', as illustrated in Figure 14.3, is commonly used to show such a programme. If the contractor has produced a programme which he has submitted to the architect or project manager or if he is willing to make his programme available to the surveyor, so much the better. Using bill rates and prices (or schedule rates and prices, if there is no bill), each section of the work is valued, the totals being broken down into monthly amounts over the period allocated to each section, as shown in Figure 14.3. The monthly amounts are then totalled and to each total is added the appropriate proportion of the Preliminaries. The final totals for each month (which together should equal the contract sum) can then be tabulated or plotted on a graph.

CONTRACT PROGRAMME CHART (Simplified)*

Operation	1	2	3	4	5	6	7	8	9	10	11	12	13	14	15	16	17	18	19	20	Totals
Demolition	3685	2000																			5685
Groundwork	1700	4625	5173	1458	3271	6916	6586														32129
Piling Nom.S/C		3100	5400	4235																	12735
Mains services connection	735				567	858															2160
In situ concrete	843	3729	4738	4740	1788	3911	4673	5609	5490	5250	5175	4234	5539								56008
Asphalt tanking and roofing Nom.S/C			1442	3420				3654	6796	1890											17190
Masonry			1870	7477	5991	5648	5517	6543				3681	7989	1928							46684
Structural/Carcassing Metal/Timber			2000	3550	4156	4553	3317	3334													20910
Cladding/Covering									2520	2373	2725	1885									9503
Waterproofing					1200	1320				6758	7096	2732									19106
Windows/Doors/Stairs						5205	6330	6520	5135	5640	5980	4451									40061
Surface finishes					1573	1406							330	1320	1500	1500	1200	1084			15522
Lifts Nom.S/C											4000	4800	9000	9000	5400	2430					36630
Electrical Services Nom.S/C													1755	3625	3625	3604	3060	3060	3060	1533	24002
Furniture/Equipment								4650	4000	3600											12243
Builders work sundries							1850	1956	2070	2090	1548										9514
Disposal systems										3112	3112	2334									8558
Piped supply systems							1124	2250	2250	2100	2000	2360	2350	562	1686	2350	662				19684
Drainage	950	1050			2458	2551	1962	1683	1683	1870	934	863									16004
Fencing					3701	1851	1851	2312	1055								4625	1849			17244
Paving															1497	998	1497	998	1953	1497	8440
Provisional Sums	50	50	50	50	50	50	50	50	50	50	50	50	50	50	50	50	2000	1558	2150	50	6458
Preliminaries	3730	2080	2751	2012	1727	1722	1792	1910	1282	1865	1724	1190	1452	1869	1889	1734	2010	2897	1944	2700	41480
Daywork and Contingencies	612	603	602	603	602	603	602	603	602	603	602	603	602	603	602	603	602	603	602	603	12050
Totals	13245	19026	24004	24575	27543	35659	33367	41401	37836	36681	34148	33284	33751	19715	15188	11502	12517	16506	9768	8234	495000

Notes: Adjustment for errors in bills of quantities distributed over monthly values. Dayworks and contingencies have been included so that the total agrees with the Contract Sum.

Figure 14.3 Use of programme chart for expenditure forecasting (*relating to contract in Appendix A, p. 336)

A graph plotted from such data is usually referred to as an 'S-curve' graph because of the shape of the graph. It indicates a build-up in the rate of expenditure over the first two or three months, an acceleration of the rate over the main part of the contract period and a gradual run-down over the final three months or so. Figure 14.4 shows a typical S-curve graph for a £500 000 contract over a fifteen-month contract period.

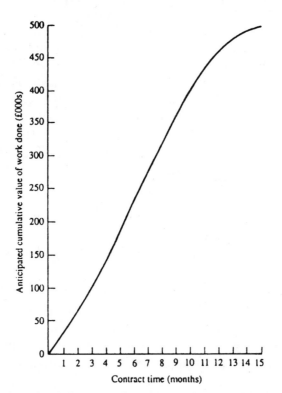

Figure 14.4 Standard S-curve of anticipated 'rate of spend'

A graph produced as described above will, of course, indicate anticipated cumulative monthly total values of work done and not a forecast rate of client's expenditure. The difference between the two is the amount of retention deductible during the contract period and the amounts of releases of retention at its end. From the amounts of monthly values of work should be deducted the proportion of value to be retained, and amounts of releases should be added. The latter may be only at the end of the contract period or may be earlier in the case of sectional completion. The adjusted amounts may then be plotted to give an S-curve showing expected rate of expenditure by the client.

There is an alternative method of producing such a graph. If the totals of interim valuations for a number of contracts are plotted and curves of 'best fit' are drawn an S-curve of approximately the same shape will result for each contract. It follows, therefore, that it is possible, within a given range of cost and over a limited range of contract times, to produce a 'standard' S-curve which can be used to predict expenditure flows for further contracts without the need to go through the detailed procedure described above.

A third method of producing an S-curve graph of value of work executed is by the use of a formula. This method was used by the Department of Health and by the Property Services Agency (now defunct).[2] It is really a development from the method described in the last paragraph because, having produced standard S-curves in the manner described, the Department of Health then proceeded to determine the underlying mathematical formula. The formula devised and used by the Department is

$$y = S\left[x + Cx^2 - Cx - \frac{1}{K}(6x^3 - 9x^2 + 3x)\right]$$

where
y = cumulative value of work executed in £
S = contract sum in £
x = proportion of contract time completed
C and K = parameters relating to contract value:

contract value £s	C	K
10,000–50,000	−0.439	5.464
50,000–100,000	−0.370	4.880
100,000–200,000	−0.295	4.360
200,000–500,000	−0.220	3.941
500,000–2m	−0.145	3.595
2m–3½m	0.010	4.000
3½m–5½m	0.110	3.980
5½m–7½m	0.159	3.780
7½m–9m	0.056	3.323
over 9m	−0.028	3.090

It will be obvious that the formula has to be used for each month's anticipated value of work done – which means fifteen times for a fifteen-month contract, etc. The mathematical work entailed presents no great problem when a computer is available, but without one the volume and tedium of the necessary calculations may deter many surveyors from using this method.

It is important, of course, that having produced a forecast of rate of expenditure, the information should be given to the client, either direct or through the architect, so that he may use it to arrange finance for the project. It may be necessary on large jobs to revise the forecast rate of expenditure at any point during the contract period for the remaining period, in the light of actual expenditure, and if this is done, again, the client should be given the revised information without delay.

A further use of the S-curve graph is to plot the totals of interim valuations as they are done, thus producing a curve of actual expenditure. If the latter is seen to be beginning to fall short of the anticipated expenditure, it will provide early warning of possible trouble. There may be a satisfactory explanation, such as a prolonged period of bad weather or a strike in the construction industry or in a related industry. If there is no obvious explanation, there may be an indication that the contractor is running into difficulties, financial or otherwise. The early warning may provide an opportunity to remedy the situation.

In such circumstances as those described in the last paragraph, the formula can be used (*a*) to determine a revised completion date; (*b*) to forecast likely expenditure in future months, based on actual expenditure in past months; and (*c*) to calculate the rate of expenditure necessary to achieve completion by the original date stated in the contract[3] (although experience has shown that this is seldom possible).

Figure 14.5 shows a graph of anticipated and actual expenditure for the contract of which particulars are given in Appendix A. All values shown are those before deduction of retention. Anticipated expenditure has been calculated using the DHS formula, using values $C = -0.220$ and $K = +3.941$. The calculated forecast of cumulative expenditure for each month is as follows:

Month	Expenditure	Month	Expenditure
1	£ 13 814	11	£308 529
2	£ 32 170	12	£341 223
3	£ 54 505	13	£372 242
4	£ 80 250	14	£401 021
5	£108 843	15	£426 994
6	£139 717	16	£449 598
7	£172 307	17	£468 265
8	£206 045	18	£482 431
9	£240 376	19	£491 531
10	£274 725	20	£495 000

The curve of actual expenditure shows monthly values before deduction of retention and exclusive of materials on site, fluctuations and claims, thus allowing an undistorted comparison

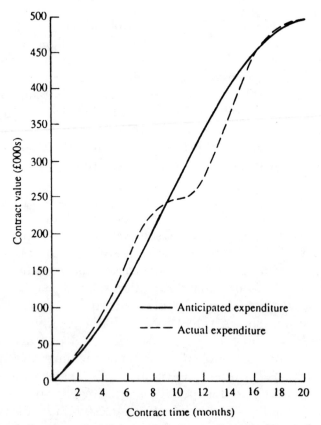

Figure 14.5 S-curve of anticipated and actual expenditure (before deduction of retention)

to be made. It will be seen that the expenditure during the first nine months was higher than forecast, but after that it dropped markedly below the forecast level. The reasons for this were: bad weather, the shut-down for Christmas 1994 and the New Year and the subsequent cessation of work on the lifts installation (see p. 209). From March onwards, the rate of production increased and by July, actual expenditure had almost caught up with the forecast value.

By plotting actual expenditure each month, it was possible to see at the December valuation that the curve was flattening out, thus giving warning of possible problems ahead. In fact, the causes were identified, as noted above, and steps were taken to remedy the situation so that the contract was eventually completed on time.

Forecasting contractor's earnings and expenditure

The same information as is produced by the forecasting techniques described above is of value to the contractor. It enables him to monitor the financial progress of each of the contracts running at any time. The S-curve of anticipated cumulative values of work done represents to the contractor his anticipated or 'programmed' earnings. The rate of client's expenditure becomes, from his point of view, actual earnings.[4] A comparison between the two provides an indication whether the financial progress of the contract is satisfactory or not.

Similarly, a comparison may be made between earnings and expenditure and a graph plotted each month showing cumulative values will indicate clearly whether a contract is making a profit or a loss. If it is desired to establish more specifically where a profit or a loss is being made, the earnings may be broken down approximately into the relevant components for ready comparison with expenditure. Gobourne gives an example of such a comparison on p. 9 of the reference last quoted.

The value of such comparisons, if carried out at least once a month, is that they enable the contractor to have early warning of a loss-making situation before the position becomes irretrievable. By pin-pointing where the losses are being made, the contractor can take action to remedy the situation. Again, prompt and up-to-date information is the key to practical use of financial analysis.

Otherwise, its value is merely historical. Thus, the contractor's surveyor who keeps the contracts manager regularly and accurately informed on the current relationship between earnings and expenditure is of considerable value to the contractor.

References

1. AQUA GROUP, *Contract Administration for Architects and Quantity Surveyors*, 7th edn, (London: Collins, 1986), p. 14.
2. HUDSON, K. W., DHSS expenditure forecasting method, *Chartered Surveyor, BQS Quarterly*, Vol. 5, No. 3, Spring 1978, p. 42.
3. COST INTELLIGENCE SERVICE, DHSS, *Health Notice HN (78) 149 Forecasting Incidences of Expenditure*, reference CON/C462/17, November 1978, pp. 2–3.
4. GOBOURNE, J., *Cost Control in the Construction Industry* (London: Butterworth & Co. Ltd, 1973), pp. 7–8.

15

Health and safety: The Construction (Design and Management) Regulations 1994

Introduction

The Construction (Design and Management) Regulations 1994[1] (the Regulations) came into being by virtue of EU Directive 92/57/EEC on the implementation of minimum safety and health requirements at temporary or mobile construction sites. The Regulations are aimed at improving the health and safety management of construction projects. They consider all involved in the construction process (clients, designers, contractors and sub-contractors) and they make provision for safety to be considered for the whole life of construction projects from inception through design and construction, maintenance, adaptation and finally, demolition.

Surveyors should have a working knowledge of the Regulations to ensure that they comply with the duties imposed upon them by the Regulations. The term 'designer' in the Regulations refers to any person or organization that prepares a design; the term 'design' includes drawings, design details, specifications and bills of quantities. Much of the documentation produced by surveyors will therefore be classified as design. Surveyors may also be involved in the management of health and safety for a construction project and could, in some situations, be appointed to act as client and/or planning supervisor (a special duty holder created by the Regulations).

This chapter aims to provide a general understanding of the nature of the Regulations, to identify the duties placed on the various parties to the construction process by the Regulations, to explain when the Regulations apply to a construction project and to consider the relationship of the Regulations to other health and safety legislation.

The development of health and safety legislation

Health and safety law began life in the UK during the Industrial Revolution, when it became clear that some legal protection was required for employees to ensure their health and safety in the workplace. Early legislation, much of which is still applicable, was generally concerned with the provision of physical measures and prescriptive rules. There are many general and specific regulations which are applicable to construction, but it is beyond the scope of this chapter to identify and explain all the legislation.

A big step forward was made with the introduction of the Health and Safety at Work, etc. Act 1974. This Act created the Health and Safety Commission, which has the power to propose health and safety regulations and to approve codes of practice, and the Health and Safety Executive, which is responsible for enforcing health and safety legislation. Unlike previous health and safety legislation, the Act is concerned with improving the attitude of employers (and employees) to health and safety by providing a broad legislative framework incorporating general duties and instilling an organized approach to safety in the workplace. The Act is *framework legislation*; that is, it is the vehicle through which new regulations, including those resulting from EU Directives, are introduced.

Breach of health and safety legislation is a criminal offence; conviction can result in a fine and/or imprisonment.

'Reasonably practicable'

Many of the duties and requirements imposed by the Regulations are governed by what is reasonably practicable. 'Reasonably practicable' was explained by Asquith, L.F. in *Edwards v National*

Coal Board [*1949*, 1 KB 704] with regard to the duty of a mine owner to support the roof of a mine:

> 'Reasonably practicable' is a narrower term than 'physically possible', and implies that a computation must be made in which quantum of risk is placed in one scale and the sacrifice involved in the measures necessary for averting the risk (whether in money, time or trouble) is placed in the other, and that, if it be shown that there is a gross disproportion between them – the risk being insignificant in relation to the sacrifice – the defendants discharge the onus upon them. Moreover, this computation falls to be made by the owner at a point of time anterior to the accident.

It is therefore the subject of a cost-benefit exercise of assessing the cost of removing or controlling a hazard against the probability or severity of harm resulting from the hazard.

The construction industry's health and safety record

The construction industry in the UK is dangerous in comparison with other industries. Legislation has developed over the years in response to the analysis of data collected on the causes of accidents on construction sites. However, it is difficult for legislation to keep pace with the continuous and rapid developments in construction and materials technology.

The incidence of fatal accidents and accidents causing serious injury is consistently high. A report of the Health and Safety Executive in 1988 (Blackspot[2]) analysed the circumstances of 739 deaths in the construction industry between 1981 and 1985. The report considered that most of the deaths could have been prevented, and stated that 'better management of sites through detailed pre-site planning with all who are to be involved in the job is needed to improve the general level of safety. This requires discussion with architects, engineers and other professional advisers, as well as main and sub-contractors, safety representatives and safety professionals. Co-ordination of the work, with particular attention to high risk activities, can reduce the overall risks'. The Construction (Design and Management) Regulations address these points; they consider the whole process of construction from initial design decisions through to maintenance and final disposal of buildings, place duties on all involved in the construction process, require the management and co-ordination of health and safety and require hazards to be identified and risks assessed.

The Regulations

The Regulations aim to ensure that health and safety is taken into account at all stages of the construction process. Construction involves many diverse activities. The employment and contractual arrangements across the industry and, indeed, for any one project, are complex and varied. The Regulations therefore focus on the management and co-ordination of activities to ensure a rational and collective approach to health and safety. They impose duties on clients, contractors and designers; they also create two health and safety management roles in the construction process; namely, the planning supervisor and the principal contractor. Joyce[3] notes, 'The planning supervisor will undoubtedly become an important source of advice to clients with regard to the Regulations'.

The Regulations have come into being in response to EU Directive 92/57/EEC on the implementation of minimum safety and health requirements at temporary or mobile construction sites which was adopted under Article 118A of the Treaty of Rome. The Regulations have been made under the Health and Safety at Work, etc. Act 1974. The Health and Safety Commission has given approval to the Approved Code of Practice L54 (ACOP) which gives advice on how to comply with the Regulations. The legal status of the ACOP is such that if it is proved in a prosecution for breach of health and safety law that the provisions of the ACOP have not been followed, then a person or organization will be found at fault (unless it can be shown that the law has been complied with in some other way).

Application of the Regulations

The Health and Safety Executive is the enforcing authority for the Regulations. The Regulations are broadly applicable to all construction projects (with some exceptions), but they always apply to work involving demolition. If the construction phase of a project is expected either to be longer than 30 days or to involve more than 500 person/days then the Health and Safety Executive requires written notification of the project. The requirements on designers are applicable to all projects, irrespective of size or duration. Where a client believes that a project is not notifiable and will involve fewer than five persons at work carrying out construction at any one time, then the majority of the Regulations do not apply. In situations where the Local Authority is the enforcing agency for the construction work, the Regulations do not apply. When there is only one designer or one contractor associated with a construction project, the Regulations regarding co-ordination of design and construction respectively do not apply. If work is carried out for a domestic client, the majority of the Regulations do not apply; however, the Regulations place responsibilities on clients who commission work in connection with a trade, business or other undertaking. They also apply to speculative residential developers as if they were clients.

The health and safety plan

The health and safety plan is the medium through which health and safety associated with the construction of a project is managed and communicated. The health and safety plan will initially be prepared during the design phase of a project, and the planning supervisor has a duty to ensure that the plan is prepared. The plan is likely to be a collaborative effort, with information being provided by the client, the designers and the planning supervisor. The plan should be included with general tender information to contractors to enable them to include in their tender for dealing with the risks, issues and requirements of the plan.

The health and safety plan for a project is not a static document; it should be continuously reviewed, updated and amended as necessary. The principal contractor has a duty to develop the health and safety plan throughout the project and to support this there is a requirement (Regulation 19) for contractors to co-operate with the principal contractor, to comply with the rules of the health and safety plan and to provide information which might justify a review of the plan. Indeed, the

Approved Code of Practice (paragraph 106) states, 'The development of the health and safety plan and its effective implementation depends upon the flow of information from the contractors to
the principal contractor'.

Information to be included at the preparation stage of the
health and safety plan is detailed in Regulation 15(3). Possible
information that could be included in a pre-tender stage health
and safety plan is listed in Appendix 2 of *A guide to managing
health and safety in construction*[5]. This list includes details of: the
nature of the project; the existing environment; existing drawings; the design (including details of inherent hazards, work
sequences, the principles of structural design and identification of
specific problems for which the contractor will have to develop
and propose risk management solutions); construction materials
(associated hazards and precautions); site set-up (e.g. access,
loading areas, traffic routes, etc.); site rules and requirements for
continuous liaison.

The health and safety plan should be developed by the principal
contractor before the construction starts and throughout the
construction phase of the project. There is no requirement for the
plan to continue after construction or for the principal contractor
to hand over the plan on completion. The principal contractor is
required by the Regulations to develop the health and safety plan.
He should incorporate within the plan: the general framework for
the management of health and safety for the project; risk
assessments prepared by contractors in compliance with the
Management of Health and Safety at Work Regulations (1992);
rules for the management of construction required for reasons of
health and safety; common arrangements (e.g. emergency procedures); details concerning co-operation, compliance with rules in
the health and safety plan, authorized persons, the issuing of
instructions and collection of information concerning health and
safety; details explaining the requirements for providing information and training; and arrangements for allowing and co-
ordinating the views of persons at work in respect of health and
safety pertaining to the project.

The importance of the health and safety plan is clear: it is the
prime tool for planning safety management for a project and all
parties are involved in its development. The plan should provide
a co-ordinated approach to the management of health and safety
on site and it should incorporate methods for dealing with

identified risks. It is likely that, when checking for compliance with the Regulations, the health and safety enforcement agencies will start by examining the health and safety plan, which could prove to be vital evidence in any decision to prosecute or, indeed, in a defence.

The health and safety file

The purpose of the health and safety file is to provide a comprehensive record of information pertaining to a building or structure to enable future persons involved in the maintenance, alteration and demolition of the building to design, plan and execute this work with due regard for health and safety. The health and safety file should be kept by the client and must be updated to take account of any changes that are made to the building or structure.

Typical information to be included in the health and safety file is listed in Appendix 5 of the Regulations and includes: record or 'as built' drawings and design criteria; general details of construction method and materials; details of equipment and maintenance facilities; specialist contractors' and suppliers' manuals giving details of operating and maintenance procedures; and details regarding the nature and location of utilities and services.

The client

Clients are required to determine if the Regulations apply to a project and, where appropriate, to appoint a planning supervisor and a principal contractor for all construction projects. The appointments are to be made as soon as practicable, and they should be reviewed, terminated and changed as necessary. Appointments must be in place throughout all projects up to completion of the construction phase. The Regulations permit the same person (or organization) to be appointed planning supervisor and principal contractor for a project (the principal contractor must be a contractor undertaking or managing construction work on the project) and they permit clients to appoint themselves as planning supervisor and/or principal contractor, providing they satisfy the requirements regarding competence and adequacy of resources.

Clients must give the planning supervisor information about the site which is relevant to the function of planning supervisor (Regulation 11). Such information is likely to include details of the existing buildings, land and associated plant. The health and safety file for a structure should be kept available by the client for inspection by any person who might need the information to comply with statutory requirements (Regulation 12). When disposing of an interest in a structure, a client must pass on the health and safety file to the relevant person or organization.

The Regulations impose a duty on clients to ensure that construction does not start until a health and safety plan for the project has been prepared (Regulation 10). It is anticipated that many clients may need to seek professional advice in making such a judgement, and it is imperative that sufficient time is allocated to principal contractors, prior to construction starting, to develop the plan. The ACOP recognizes that the use of some procurement methods will mean that there is often an overlap between the design, planning and construction phases of a project. The health and safety plan therefore needs only to be developed before construction may start, so far as the general framework for health and safety management of the project is concerned, together with those work packages which can reasonably be developed before the construction phase begins.

The duties imposed upon clients by the Regulations are onerous. They are permitted to appoint an agent to act as client in respect of the Regulations and therefore to undertake the duties imposed upon clients by the Regulations. This would be advisable action for clients with little or no experience of the construction process. Additionally, where there is more than one client for a project, one of the clients (or an agent) can elect to take the responsibility of the client under the Regulations.

The planning supervisor

Primarily, the planning supervisor has a co-ordination and advisory role. Duties imposed on the planning supervisor include: ensuring that designers comply with the Regulations and co-operate with other designers; being in a position to give advice to clients and contractors with regard to competency and adequacy of resources associated with engaging consultants, contractors and sub-contractors; ensuring that the health and safety file is prepared; reviewing, amending and adding to information in the health and safety file; and ensuring that the health and safety file is delivered to the client on completion of the construction work. The planning supervisor can be a company, a partnership or an individual.

The planning supervisor is required to give written notice of the project to the local office of the Health and Safety Executive (Regulation 7). The notice is to be given as soon after the appointment of the planning supervisor as is reasonably practicable. Information to be included in the notice includes: the address of the site; the names and addresses of the client, planning supervisor, principal contractor and any contractors chosen; the type of project; a declaration of the appointment of the planning supervisor, signed by or on behalf of the planning supervisor; a similar declaration for the principal contractor; the planned date for the start of construction; the anticipated construction duration; the estimated numbers of people at work on the site and the planned number of contractors on the site. The Executive has produced Form 10(rev) which can be used for giving notification (use of this form is not mandatory). Any information which is not available at the time the notice is given should be forwarded to the Health and Safety Executive as soon as it becomes available. Where the project is for a domestic client, the contractor is required to give written notice to the Executive before construction work starts.

Designers

The definition of 'designer' under the Regulations is wide, and includes activities that would perhaps traditionally not be thought of as design. Design includes drawings, design details, specifications and bills of quantities. The term 'designer' refers to any person who prepares (or arranges for persons under his or her control to prepare) a design. Thus surveyors, when performing some of their services, may be classified as designers within the meaning of the Regulations.

Designers must ensure that clients are aware of their (the clients') duties under the Regulations (Regulation 13(1)). They must include in their design adequate information about any aspect of the project that might affect the health and safety of those constructing and maintaining it. Designers are required to co-operate with the planning supervisor and other designers insofar as such co-operation will enable each of them to comply with statutory provisions.

Design work should be undertaken with a view to avoiding foreseeable risks to the health and safety of those carrying out associated construction, repairs and maintenance (and any person who might be affected by the work of such persons), combating risks to their health and safety and prioritizing health and safety measures which will protect all persons. Management of risks to health and safety is a key duty, and the onus is on designers to undertake risk assessment and to make design decisions which include consideration of the health and safety of those who will be constructing, maintaining, altering and demolishing the structures designed. It is not the intention of the Regulations that all risks should be avoided; decisions should be made by designers within the context of reasonable practicability. Reasonable practicability should be considered in terms of the costs of taking a course of action (measured in terms of finance, fitness for purpose, aesthetics, buildability and environmental impact) and the benefits that flow from that action.

The hierarchy of dealing with risks gives precedence to designing to avoid risks. If this is not possible, the causes of risks should be tackled at source and, failing this, the effects of risks should be reduced and controlled by protecting those persons whose health and safety might be affected by them (see Figure 15.1). The spirit of the Regulations in this respect is for designers to include in their decision-making process the effect that their design has on those who will be constructing and maintaining it. It has been recognized that decisions made during the design process can have an impact on health and safety. It is important to appreciate that this is about making judgements and exercising management. As with design, risk assessment and management should be a continuous process; health and safety issues should fall among all the other variables within the design decision process and should not be considered in isolation. The Regulations do not require designers to keep written records of health and safety considerations made during the design process, but there may be benefits associated with keeping records – for

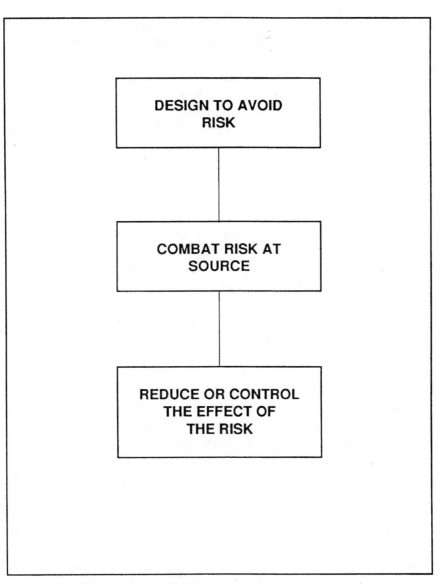

Figure 15.1 Hierarchy of risk management

example, to help explain retrospectively the reasons behind a decision or course of action.

The duties on designers in respect of preparing designs are limited to the extent of what it is reasonable for a designer to do at the time the design is prepared. Designers have a duty to pass information to the planning supervisor so that it can be included in the health and safety plan and the health and safety file, and to provide information on health and safety aspects and other issues of their designs to those who may need it.

A practical understanding of the processes and techniques of risk assessment, together with an awareness of the construction implications of designs, is essential if designers are to make sound professional judgements and discharge their duties under the Regulations effectively. As with all questions of professional competence, designers must be aware of their own limitations and be able to recognize when specialist technical advice should be sought.

In situations where the conditions of a building contract require a contractor or sub-contractor to undertake some of the design work for a project, the Regulations, insofar as they apply to designers, will also apply to these contractors with regard to the design element of their contracts.

The principal contractor

The principal contractor is generally responsible for the management of health and safety during the construction phase of a project. This will include co-ordination of all work with regard to health and safety. The Regulations impose duties on the principal contractor to ensure co-operation between all contractors, to develop and implement the health and safety plan, to ensure that the health and safety plan sets out the arrangements for health and safety management, to ensure that all contractors and employees comply with rules in the health and safety plan for the project, to ensure that only authorized persons are permitted on the site, to provide the planning supervisor with information for the health and safety file, to monitor health and safety performance, to provide opportunity for health and safety discussion with all persons at work on the project and to ensure that all contractors have the necessary information and training in respect of health and safety. The role is generally about ensuring an integrated approach to health and safety on site.

The Regulations give the principal contractor the right to give directions to any contractor to enable compliance with these duties. The principal contractor can also lay down rules, for the purpose of health and safety, for the management of construction work within the health and safety plan. The principal contractor must display on site (Regulation 16(1)(d)) the information provided in the notification to the Health and Safety Executive.

The risk assessments prepared by other contractors should be evaluated by the principal contractor. The objective of the evaluation should be to ensure that all risks are considered in the overall management of health and safety and that a co-ordinated approach to risk management, which takes into account the inter-relationship of different contractors' operations and their associated risks, is adopted. The principal contractor must be satisfied that the risk assessments and method statements prepared by contractors are adequate and compatible with the health and safety plan.

Communication is essential in ensuring that the health and safety plan is effectively implemented. The principal contractor must clearly communicate the requirements of the health and safety plan and ensure that arrangements for health and safety are coherently disseminated throughout the site on a regular basis.

Where a principal contractor is involved in the selection and appointment of works contractors or sub-contractors, the principal contractor should have regard to ensuring that these contractors have sufficient information regarding the health and safety requirements of the project and that they have allowed sufficient and appropriate resources in their tenders.

Contractors

The principal contractor's co-ordination role does not relieve individual contractors of their legal duties. The Regulations impose requirements and prohibitions on *all* contractors. They are required to co-operate with the principal contractor, provide information to the principal contractor which might affect the health and safety of persons at work on the project, comply with directions given by the principal contractor under the Regulations, comply with rules in the health and safety plan, provide information in accordance with the Reporting of Injuries, Diseases and Dangerous Occurrences Regulations 1985, and provide the principal contractor with information pertinent to the health and safety file. Contractors should not permit employees to work on site unless they have the name of the planning supervisor, the name of the principal contractor, and relevant parts of the health and safety plan.

The Management of Health and Safety at Work Regulations 1992 require Employers to assess the risks to health and safety arising out of their undertakings. Contractors are required to provide these risk assessments, together with details of the arrangements for health and safety resulting from the risk assessments, to the principal contractor. This enables the latter to develop and co-ordinate an integrated approach to health and safety management for a project.

Tender procedures

Effective contractor selection is very important in the management process. Tender documentation should be set out to enable judgements to be made about the adequacy of resources allowed in tenders and to prompt contractors to provide information confirming their competency. Conversely, documentation should contain sufficient information to make contractors aware of the health and safety requirements for a project and thus enable them to make adequate financial and programme provision. Tender pre-qualification procedures will help to ensure that only competent contractors are invited to tender (see Chapter 4, *Tendering methods and procedures*).

Risk assessment

Risk assessment is at the heart of health and safety management. The purpose of risk assessment is to help designers to make informed design decisions. It is about identifying hazards arising from a design solution, assessing the likelihood that harm will occur from the hazard and the probable severity of the harm caused by a hazard occurring. Common sense dictates that priority should be given to managing those risks where the severity of harm (the consequence) and the likelihood of occurrence (the frequency) are high. An approach to risk assessment is shown in Figure 15.2.

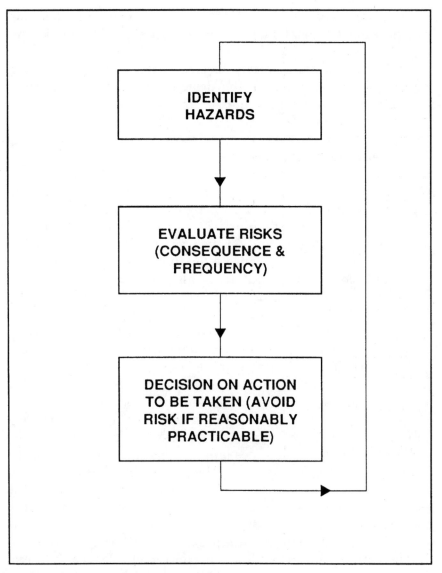

Figure 15.2 Approach to risk assessment

Hazard identification involves a systematic analysis and review of the design. Identification can be guided by a knowledge of published accident statistics (such as those published by the HSC) which will help identify where and how construction accidents are likely to occur. Risk evaluation can be undertaken using a combination of qualitative and quantitative methods.

Even crude assessments of risk are useful in aiding decision-making, and personal knowledge, experience and 'gut feeling' should not be ignored when evaluating risk. The risk assessments are essential in making rational decisions about the design and in highlighting those areas of the design where contractors need more information regarding health and safety issues.

Competence and resources

There is a duty on clients to appoint persons as planning supervisors only if they are reasonably satisfied that they are competent to perform the functions required of planning supervisors under the Regulations for the project and that they have, or will allocate, sufficient resources for the project to comply with the Regulations. The Regulations also impose a duty on any person arranging for a designer to design or a contractor to construct to be reasonably satisfied of competence and resources in performing these functions in accordance with the Regulations for the project. To ensure reasonable satisfaction, some checks will be necessary. For small or low-risk projects, it may be a simple case of asking questions on a face-to-face basis to ensure adequacy of competence and resources. For larger or more complex projects, a more formal approach may be appropriate – for example, giving written questions and requiring both written and oral responses. The types of questions that should be asked will depend on the particulars of the project, but in all cases they must be designed to obtain information that will enable the assessor to make a judgement about the adequacy of competence and resources. These questions could be included as part of the normal pre-selection procedures associated with the appointment of consultants, contractors and sub-contractors. Typical information requested might include:

1 Membership of relevant professional associations.
2 Knowledge of construction related to a particular project.
3 Knowledge of health and safety issues and requirements.
4 Ability to co-ordinate their own work with that of others.
5 Availability and suitability of human resources.
6 The time allowed to perform the required functions.
7 The availability and suitability of technical facilities.

8   The management systems that will be used.
9   The proposed methods of dealing with risks identified in the health and safety plan.
10 Arrangements for managing health and safety.
11 Arrangements for monitoring compliance with health and safety legislation.

The Regulations in respect of competence and adequate resources require reasonable satisfaction at the time of appointment or arrangement. There is no requirement for competence and resources to be continually monitored once appointments or arrangements have been made.

Relationship of the Regulations with other health and safety legislation

The Regulations apply simultaneously with other health and safety legislation. Many of the health and safety regulations relating to construction are specific in nature. They prescribe actions to be taken or provisions to be made to control or avoid specific risks. The Health and Safety at Work, etc. Act improved on this situation. The Act gave emphasis to individuals and their duties. The Management of Health and Safety at Work Regulations 1992 were introduced to improve health and safety management and to make more explicit the requirements of the Health and Safety at Work, etc. Act insofar as they place duties on Employers. They are applicable to all Employers and they apply to all work on construction sites.

The Management of Health and Safety at Work Regulations 1992 generally require Employers to take an active role in managing health and safety in the workplace. Employers are required to assess risks, make appropriate arrangements for implementing measures in the light of risk assessment, provide health surveillance for employees, appoint competent persons to help devise and apply health and safety measures, develop emergency procedures, provide information and ensure adequate training for employees on health and safety, co-operate and co-ordinate activities with other Employers sharing the same workplace, and provide temporary workers with some health and safety information.

The Construction (Design and Management) Regulations 1994 extend the duties imposed by the Management of Health and Safety at Work Regulations 1992. For example, designers have a duty under the 1992 Regulations to assess the risks arising out of their business; the 1994 Construction Regulations explicitly extend this duty to considering the risks in constructing a design. The Construction (Design and Management) Regulations address the specific characteristics involved in managing safety in construction. They take account of the construction process, the parties involved in the process, the complex contractual arrangements and the variety of different procurement and employment arrangements.

References

1. HSE, *Managing Construction for Health and Safety: Construction (Design and Management) Regulations 1994: Approved Code of Practice* (Sudbury: HSE Books, 1995).
2. HSE, *Blackspot Construction* (London: HMSO, 1988).
3. JOYCE, R., *The CDM Regulations Explained* (London: Thomas Telford Publications, 1995).

Bibliography

1. HSE, *Managing Construction for Health and Safety: Construction (Design and Management) Regulations 1994: Approved Code of Practice* (Sudbury: HSE Books, 1995).
2. JOYCE, R., *The CDM Regulations Explained* (London: Thomas Telford Publications, 1995).
3. BARRETT, B. and HOWELLS, R., *Occupational Health and Safety Law* (2nd edition) (London: Pitman Publishing, 1995).
4. HSE, *Designing for Health and Safety in Construction* (Suffolk: HSE Books, 1995).
5. HSE, *A Guide to Managing Health and Safety in Construction* (Suffolk: HSE Books, 1995).
6. The Institution of Civil Engineers, *The Management of Health and Safety in Civil Engineering* (London: Thomas Telford Publications, 1995).

16

Indemnity and insurance

Introduction

The risks of damage and injury associated with construction projects are often high. Insurance provision is therefore an essential requirement in construction contracts to protect the interests of those involved. It is, without question, necessary for those engaged in advising clients and administering contracts to have a basic knowledge of the principles of indemnity and insurance, together with an understanding of the provisions and operation of the associated clauses in standard forms of building contract. This knowledge and understanding will help to ensure that:

1   Clients are informed about the responsibilities for insuring risks under a contract, thus enabling them to comply with any obligations imposed upon them by the contract conditions and to arrange additional cover where necessary.
2   The most suitable insurance clauses (from the various alternative standard clauses) for individual projects are used and an adequate level of financial cover is stipulated within contract conditions.
3   The insurance policies put in place for a contract comply with the requirements for insurance cover in the contract conditions.
4   The correct procedures are adopted in the event of an insurance claim associated with loss or damage.

This chapter provides a basic introduction to indemnity and insurance associated with building contracts. In particular, it focuses on the provisions of the Standard Form of Building Contract 1980 Edition, Private With Quantities version,[1] and references to contract clauses in this chapter relate to this contract.

Some general principles of insurance

There are two basic types of insurance:

1 *Liability insurance,* providing financial cover for legal liabilities owed to others.
2 *Loss insurance,* providing cover for losses which fall directly on the insured parties.

A contract of insurance is a legal contract, and therefore the general criteria relating to the existence of a contract pertain. To take out insurance cover, a person or organization must have an insurable interest (e.g. ownership of a house). The principle of utmost good faith requires the insured to disclose to the insurers all material information in respect of the risks being insured. The principle of indemnity means that insurance will put the insured back into the position they would have been in had the risk not occurred (i.e. no loss and no gain). *Subrogation* is the insurers' right to pursue claims that the insured may have against third parties in respect of the insurers' indemnification.

'Standard' insurance policies do not exist. It is therefore essential to check that the insurance provided by a policy gives the protection required. This can be difficult, as many of the terms used in insurance policies can be open to different interpretations; the plethora of case law on this subject highlights the problem. It should also be borne in mind that the insurance market is not static; insurance policies and products are continually being updated to take account of new technologies and new risks – policies must therefore be carefully scrutinized.

Indemnity

Clause 20 of JCT 80 places obligations upon contractors to indemnify Employers against the consequences of injury to persons and property.

Clause 20.1 requires contractors to indemnify Employers in the event of personal injury or death to any person arising out of, or in the course of, or caused by, the carrying out of the Works. A contractor's liability under clause 20.1 is for any expense, liability, loss, claim or proceedings against an Employer arising under any statute or at common law. A contractor's liability does not, however, extend to injury or death arising as a result of an act or neglect on the part of an Employer or of persons for whom an Employer is responsible (including persons directly employed as defined by clause 29).

Clause 20.2 requires contractors to indemnify Employers in respect of any expense, liability, loss, claim or proceedings arising from injury or damage to property, real or personal. Injury to property could include situations where, although there may not have been physical damage, enjoyment of a right, for instance, may have been affected – for example, an easement or a loss of light caused as a result of construction activity. This indemnity is limited to injury or damage arising out of, or in the course of, or by reason of, the carrying out of the Works. Indemnity is further limited to injury or damage to property due to negligence, breach of statutory duty, omission or default of 'the Contractor, his servants or agents or of any person employed or engaged upon or in connection with the Works or any part thereof, his servants or agents or of any other person who may properly be on the site . . . his servants or agents'.

The contractor is not, however, responsible for injury or damage to property caused by the 'Employer or any person employed, engaged or authorised by him or by any local authority or statutory undertaker executing work solely in pursuance of its statutory rights or obligations'. Where a statutory body is acting as a domestic or nominated sub-contractor, then the contractor will be liable for injury or damage caused by them under clause 20.2. 'Property real or personal' excludes the Works, any work executed, and materials on site. Damage to the Works etc., is covered by the insurance provisions of clause 22; clause 20.2 deals with indemnifying the Employer against damage to third party property.

Where work is to existing structures (for example, repair, alteration or refurbishment projects) and clause 22C applies with regard to insurance of the Works, the indemnity required by clause 20.2 excludes injury and damage to the existing structures and their contents resulting from Specified Perils. (The definition of Specified Perils is given in clause 1.3 as 'fire, lightning, explosion, storm, tempest, flood, bursting or overflowing of water tanks, apparatus or pipes, earthquake, aircraft and other aerial devices or articles dropped therefrom, riot and civil commotion, but excluding Excepted Risks'.) This is dealt with in clause 22C.1. However, indemnity is required for injury or damage caused other than by a Specified Peril.

Insurance against injury to persons and property

The requirements

Clause 21 of JCT 80 provides a financial 'backbone' to the indemnities given to the Employer under clause 20. It places a responsibility upon contractors to effect insurance cover against their liabilities under the contract. Insurance is described by Madge[2] as 'a risk-spreading mechanism, spreading the losses and liabilities of the few amongst the many'. The cost of insurance will, to a large extent, be related to the types of risks being taken; the more claims and awards there are across the industry, the higher will be the cost of insurance. This cost is passed on to the clients of the construction industry through the contracts awarded.

The insurances required by clause 21 are to cover personal injury and death to third parties and damage to property (other than the Works). Clause 21 is an insurance clause, as opposed to an indemnity clause; it places a contractual obligation on contractors to arrange insurance. It does not relieve contractors of their obligation to indemnify the Employer under clause 20 of JCT 80.

Clause 21.1 requires insurance for personal injury or death to a contractor's employees to be in accordance with the Employer's Liability (Compulsory Insurance) Act 1969. This insurance cover is a statutory requirement, and therefore the clause acts as a reminder to contractors of their statutory obligations. Other insurance required by the clause is to be sufficient to cover all other liabilities imposed upon a contractor by clauses 20.1 and 20.2. The financial cover provided must be not less than the sum stated in the Appendix to the contract 'for any one occurrence or series of occurrences arising out of one event'. The insurance cover is required to indemnify the Employer as well as the contractor. It is not practicable for Employers to require unlimited indemnity; insurance companies do not provide this cover. The minimum amount of insurance cover required must therefore be established by the Employer. The amount will depend upon the type, size, location, etc. of the project. Practice Note 22[3] advises that the amount stated must be realistic and, if necessary, advice should be sought from insurance experts in establishing the figure.

Documentary evidence

Clause 21.1.2 enables Employers to elicit documentary evidence from contractors to the effect that insurances required in respect of third party cover have been put in place and are being maintained. The procedures required by the contract are for documents to be sent to the architect (upon the request of the Employer) for inspection by the Employer. It is common, in practice, for Employers to request the architect, or design team, to check that these documents are in accordance with the requirements of the contract; indeed, where Employers are unfamiliar with construction contracts, the design team should advise them of their contractual right to request evidence in order to protect their interests.

Insurance is a complex field, and checking that the provisions of individual insurance policies comply with the requirements of building contracts can be fraught with dangers. Faced with such a request, it is important to take the correct course of action. This will depend upon whether an individual or firm has the necessary insurance expertise in each particular situation (some situations may be more straightforward than others) to verify the adequacy of cover, or if specialist advice is required. Bearing in mind the potentially large sums that may be involved in the event of an insurance claim, such decisions must be made carefully.

Length of cover

The contract is not explicit with regard to the length of time for which this insurance should be maintained. Practice Note 22[4] gives some guidance in that insurance taken out in compliance with clause 21.1 should not be terminated 'until at least the expiry of the Defects Liability Period or the date of issue of the Certificate of Completion of Making Good Defects whichever is the later'. This view is supported by Madge[5]: 'such insurances ought to remain in force at least to the end of the defects liability period'.

Default by a contractor in effecting insurance cover

If a contractor defaults in taking out or maintaining the insurance prescribed by the contract, there is a remedy for the Employer. Employers are empowered, in such circumstances, to take out insurance against any liability or expense which could arise as a result of such a default. The cost of this insurance can either be deducted from money due to the contractor from the Employer or, alternatively, can be recovered by the Employer from the contractor as a debt. It should be noted that, in these circumstances, Employers would only insure their own liability – they are not required to take out cover to protect a contractor against liability claims (in fact, it would be difficult for an Employer to arrange such cover).

Excepted risks

Contractors are not required to indemnify Employers or to insure against injury, death or damage resulting from Excepted Risks. Excepted Risks are defined in clause 1.3 of JCT 80 and include nuclear perils and the effects of sonic or supersonic aerial devices. These risks are generally excluded from the cover provided by insurance policies and hence their exclusion from the contractual requirements to insure. Liability for injury or damage resulting from these risks is generally covered by other mechanisms, such as nuclear legislation (where liability is that of licensed nuclear operators) and the UK government's undertaking to compensate for damage arising out of sonic bangs from Concorde.

Damage to property not covered by a contractor's indemnity

Indemnity is provided by a contractor under clause 20.2 only with regard to injury or damage to property due to negligence, breach of statutory duty, omission or default on the part of the contractor (or those for whom that contractor is responsible). Clause 21 requires contractors to arrange insurance to support this indemnity.

The situation could arise, however, whereby property is damaged as a consequence of construction work but not as a result of negligence, etc. on the part of the contractor. An Employer could be held liable for such damage. Typical examples of situations resulting in such damage include subsidence, heave and the lowering of ground water. JCT 80 provides an optional provision in clause 21.2 by which the liability of the Employer in such circumstances can be protected by insurance cover. This clause does not require contractors to indemnify Employers against risks, but obligates them to arrange insurance on behalf of an Employer to cover the Employer's liability.

If it is anticipated that this insurance will be required, a statement should be included in the Appendix to the Conditions of Contract to this effect (i.e. that insurance may be required) together with the amount of indemnity required. Decisions regarding whether this insurance will be necessary in a particular circumstance and, if so, the amount of insurance cover required will depend upon the particular circumstances of the Employer, the project and the site. If insurance is required, the architect must instruct the contractor to take out a Joint Names Policy for the amount of indemnity stated in the Appendix to the Conditions of Contract. The policy is required to indemnify the Employer (although it is in joint names) for any expense, liability, loss, claim or proceedings against the Employer resulting from injury or damage to property (excluding the Works and materials on site) 'caused by collapse, subsidence, heave, vibration, weakening or removal of support or lowering of ground water' as a result of the construction work. Injury or damage which is either:

(i)   a contractor's liability under clause 20.2; or
(ii)   caused by design error; or
(iii) an inevitable consequence of the construction work; or
(iv)   is the responsibility of the Employer to insure under clause 22C.1 (i.e. damage caused by Specified Perils to existing structures); or
(v)   arises from war risks or Excepted Risks,

is excluded from the insurance requirements.

The cost of this insurance should be added to the Contract Sum. The insurers used by a contractor to provide the cover required by this clause must be approved by the Employer. Policies and premium receipts are to be sent by the contractor to the architect for depositing with the Employer.

If a contractor defaults in taking out or maintaining the insurance cover as instructed, Employers are given the contractual right to take out such cover as has not been effected and recover the cost from the contractor as described earlier.

Insurance of the works

Clauses 22, 22A, 22B and 22C of JCT 80 deal with the provisions for the insurance of the Works. The general obligation is that All Risks insurance must be provided, giving cover for physical loss or damage to work executed and site materials (but with some exclusions).

Clauses 22A, 22B and 22C are the three alternative clauses which provide the mechanisms for insurance of the Works. Only one of these clauses will be applicable for any one contract. Clause 22A covers the insurance requirements for new buildings where the contractor is made responsible for arranging the insurance; clause 22B is for new buildings for which the Employer is responsible for arranging the insurance; and clause 22C should be used where work is in, on, or to, existing structures – it requires insurance of the existing structure and contents and insurance of the Works to be arranged by the Employer. The applicable clause for a contract must be stated in the Appendix to the Conditions of Contract.

The policy of insurance of the Works must be in joint names to ensure that protection is provided to the parties against subrogation. This protection is to extend to nominated sub-contractors and domestic sub-contractors (but not for loss or damage to existing structures referred to in clauses 22C.1 and 22C.3 in respect of domestic sub-contractors) in respect of loss caused by the Specified Perils only. This protection can be provided either by naming these sub-contractors or by a waiver in the insurance policy of the right of subrogation which an insurer may otherwise have against a sub-contractor. Sub-contractors will have to arrange their own insurance for loss or damage caused other than by Specified Perils.

All Risks insurance must be maintained up to the date of issue of the certificate of Practical Completion or the date of determination of the employment of the contractor, whichever is the earlier. It is essential at this point to ensure that clients are aware of the date on which insurance of the Works ceases, to enable them to arrange insurance cover for the buildings to take effect on the cessation of the Works insurance.

The insurance requirements of clauses 22, 22A, 22B and 22C do not replace the obligations placed upon contractors by clause 2.1 to carry out and complete the Works. However, while the risk may lie with them, contractors maintain the right to take legal action against third parties who are liable for loss or damage to the Works. It is in both the contractor's and the Employer's interests that insurance is taken out in joint names. This ensures that a contractor's financial exposure in the event of loss or damage is covered, which provides a safeguard for both parties. Employers have an insurable interest in the Works by virtue of interim payment procedures. Joint insurance relieves Employers of any liability they may have to an insurance company for any responsibility they may have for loss or damage caused to the Works.

Clause 22A – All Risks insurance of the Works by the contractor – New buildings

Clause 22A should be used for projects involving the construction of new buildings and in situations where the Employer does not have (or does not wish to arrange) insurance cover for the work in progress. Traditionally, it was the contractor's obligation to arrange Works insurance, but there has been a move away from this situation by Employers who gain benefit from arranging insurance themselves (for example, Employers who are frequent developers might be able to purchase insurance at a competitive price). In such circumstances, clause 22B should be used (but normally should only be contemplated at the express wish of an Employer).

Clause 22A requires contractors to take out a Joint Names Policy for All Risks insurance to cover the full reinstatement value of the Works. If a percentage to cover professional fees associated with the reinstatement of the Works is stated in the Appendix to the Conditions of Contract, insurance cover should include this percentage. The policy must be maintained until the earlier of either:

(i)   the date of issue of the certificate of Practical Completion; or

(ii) the date of determination of the employment of the contractor.

There are two alternative ways that cover may be provided by the contractor, either:

(i)   a bespoke policy for the project, placed with insurers approved by the Employer (clause 22A.2). The contractor is required to send the insurance policy and premium receipts to the architect to be deposited with the Employer; or

(ii) if the contractor has an annual insurance policy which provides All Risks insurance then it may be used, provided it complies with the requirements of the contract concerning the provision of All Risks insurance (clause 22A.3). The contractor must make the policy documentation available for inspection by the Employer if requested. The renewal date of annual policies must be stated in the Appendix to the contract, thus enabling the continuity of cover to be checked by the Employer at the appropriate times.

If a contractor fails to take out or maintain insurance in accordance with the requirements of the contract, the Employer is empowered to take out Joint Names cover in respect of the default. The cost of this insurance can either be deducted by the Employer from sums owing or recovered from the contractor as a debt.

If loss or damage occurs as a result of an insured risk, the following action should be taken:

1   The contractor must give written notice to the architect and the Employer, providing details of the extent, location and nature of the loss or damage.

2   No account should be taken of the loss or damage when calculating payments due to the contractor.

3   Once the insurers have finished their inspections of the loss or damage the contractor should restore, replace and repair as necessary and proceed with the Works. It is advisable to

obtain express notice from the insurers that they have finished inspections before starting the restoration of damage.

4 The contractor must authorize the insurers to pay all sums due under the insurance claim to the Employer.

5 The contractor is paid for the reinstatement work by the Employer in instalments as certified by the architect. These certificates should be issued at the normal period of interim certification. The contractor is entitled to receive the full sums received from the insurers (excluding sums for professional fees), but is not entitled to receive any sums beyond those received under the insurance for reinstatement.

Clause 22B – All Risks insurance of the Works by the Employer – New buildings

If clause 22B is used, All Risks insurance is to be arranged by the Employer in Joint Names. The clause requires the Employer to provide documentary evidence of the insurance when reasonably required to do so by the contractor. If the Employer defaults in taking out or maintaining insurance, the contractor can take out Joint Names cover in respect of the default and the cost is added to the Contract Sum.

The procedures in the event of loss or damage under clause 22B are similar to those prescribed in clause 22A, except with regard to payment to the contractor. Under clause 22B reinstatement works are treated as though they were a variation required by the architect under clause 13.2. Payment is not therefore governed by the limit of the monies received from an insurance claim.

Clause 22C – Insurance of existing structures and insurance of the Works in, or extensions to, existing structures

This clause is to be used when construction involves work to an existing building (for example, refurbishment, repair, fitting-out, extensions, etc.). There are two separate issues to deal with here:

1 Insurance of the existing structure and its contents.
2 Insurance of the Works.

The Employer is required to arrange a Joint Names Policy in respect of the existing structure and its contents (clause 22C.1). The insurance must provide cover for the full cost of reinstatement, repair and replacement of loss or damage caused by the Specified Perils. This insurance must be maintained until either the date of issue of the certificate of Practical Completion or the date of determination of the employment of the contractor, whichever is the earlier. In the event of a claim, the contractor is required to instruct insurers to pay all sums to the Employer.

The requirements for insurance of the Works, and the procedures and provisions with regard to the occurrence of loss or damage, are similar to those under clause 22B (as described above). However, there is an additional provision in clause 22C.4 whereby a right is given to either party (if just and equitable) to determine the employment of the contractor within 28 days of the loss or damage.

The remedies for the contractor, in the event of an Employer failing to take out and maintain insurance as required, are the same as those in clause 22B.

General note on Clauses 22A, 22B and 22C

There may be circumstances where it is impossible to arrange the complete insurance cover prescribed by the contract for a project (for example, in areas where there is a high risk of flooding there may be exclusions relating to flood damage). In such circumstances, the footnotes in the contract suggest that arrangements should be agreed between the parties and the Contract Conditions amended accordingly.

Clause 22D – Insurance against loss of liquidated damages

Clause 22D provides the option for insurance cover in respect of an Employer's loss of liquidated damages. Such a loss would arise if an extension of time is granted under clause 25.3 for delays to the Works resulting from loss or damage to the Works caused by a Specified Peril. If this insurance is required, the contractor must arrange it, and the cost of the cover is added to the Contract

Sum. If the insurance *may be* required, a statement to this effect should be made in the Appendix to the Contract Conditions.

References

1. JCT, *Standard Form of Building Contract, 1980 Edition, Private With Quantities* (London: RIBA Publications Ltd, 1994).
2. MADGE, P., *A guide to the indemnity and insurance aspects of building contracts* (London: RIBA Publications Ltd, 1985), p. 41.
3. JCT, *Practice Note 22 and Guide to the Amendments to the Insurance and Related Liability Provisions 1986* (London: RIBA Publications Ltd, 1986).
4. ibid., p. 34.
5. MADGE, P., *A guide to the indemnity and insurance aspects of building contracts* (London: RIBA Publications Ltd, 1985), p. 43.

Bibliography

1. JCT, *Standard Form of Building Contract, 1980 Edition, Private With Quantities* (London: RIBA Publications Ltd, 1994).
2. MADGE, P., *A guide to the indemnity and insurance aspects of building contracts* (London: RIBA Publications Ltd, 1985).
3. JCT, *Practice Note 22 and Guide to the Amendments to the Insurance and Related Liability Provisions 1986* (London: RIBA Publications Ltd, 1986).
4. EAGLESTONE, F., *Insurance under the JCT Forms* (London: Collins Professional and Technical Books, 1985).
5. EAGLESTONE, F. N., *Insurance for the Construction Industry* (London: George Godwin Ltd, 1979).
6. RICS, *Introductory Guidance to Insurance under JCT Contracts* (London: RICS Books, 1991).

17

Taxation

Introduction

Clients' accountants sometimes request information concerning the value of machinery and plant included in interim payments to contractors. Why do they require this information, and what is their interpretation of the terms 'machinery' and 'plant'? Again, clients often seek advice from their professional consultants with regard to the VAT they are being charged by contractors. In which situations should VAT be charged for construction work, and how much should be charged?

This chapter examines the basic principles of capital allowances and VAT and considers the specific application of these principles in relation to construction and buildings.

Taxation is often considered to be a complex specialization and an area in which only taxation experts should be involved. While there may be some truth in this view, it is, nevertheless, essential for surveyors to have a basic understanding of taxation in the context of construction and buildings. This basic understanding should be sufficient to enable them to advise and help clients or Employers with regard to:

(i) preparing calculations to support claims for capital allowances;
(ii) using tax planning to assist in making effective design decisions;
(iii) ensuring that VAT is charged correctly.

This chapter identifies the importance of surveyors understanding taxation in order to ensure that clients' interests are properly protected. Many surveying practices have, over recent years, expanded their remit to embrace specialist financial services, including taxation advice.

Capital allowances

Capital allowances are available to taxpayers against capital expenditure on certain types of fixed assets. Capital allowances reduce the tax liability of a business. There are a number of different categories of qualifying expenditure which can attract capital allowances. These categories include:

(i)   machinery and plant;
(ii)   industrial buildings;
(iii)   agricultural buildings;
(iv)   hotels;
(v)   buildings in enterprise zones;
(vi)   patent rights;
(vii)   know-how;
(viii) scientific research;
(ix)   mineral extraction.

The principal legislation governing capital allowances is the Capital Allowances Act 1990. There are several general requirements that must be satisfied if expenditure is to be eligible for capital allowances. First, it must be expenditure incurred by the person or organisation claiming the allowance; secondly, the expenditure must have arisen in the course of a trade; thirdly, the expenditure must have occurred in a chargeable period; and fourthly, the asset must be owned by the person or organization claiming the allowance.

Expenditure can qualify for relief even if it is funded by a loan or hire purchase agreement. In both these situations, only the cost of the asset is allowable (although interest and hire purchase charges are allowed separately as a business expense). However, in the case of assets provided under a lease agreement, it is the lessor who is entitled to the capital allowances.

Capital allowances are available in respect of chargeable periods. Allowances are first given in the chargeable period in which expenditure is incurred; this is known as the *basis period*.

For a company, the chargeable period will be the accounting period.

The allowances most often available in respect of property are machinery and plant, industrial buildings, hotels, and buildings in enterprise zones. These allowances will be considered in further detail.

Machinery and plant

One of the main difficulties in claiming allowances for machinery and plant stems from the problem of deciding which assets are included in each of the terms 'machinery' and 'plant'. There is no definition of these terms within the taxation statutes. However, the Finance Act 1994 has gone some way towards clarifying the distinction between plant and buildings. 'Machinery' is reasonably straightforward; 'plant', however, is more complex, and, as a result, there is much case law concerning its definition. The key distinction between what can and cannot be claimed as plant is that apparatus used to carry on a business is 'plant', whereas the setting in which the business is carried on is not plant. Loose items of equipment do not generally cause problems in classification; it is those items of machinery and plant that are fixtures within, or form part of, a building or structure that give rise to difficulties.

Expenditure in respect of buildings, fixed structures and interests in land is not classified as machinery and plant. For buildings, this includes walls; floors; ceilings; doors; windows; stairs; mains services; systems of water, electricity and gas; waste disposal systems; sewerage and drainage systems; shafts or other structures in which lifts, hoists, escalators and moving walkways are installed; and fire safety systems. However, capital allowances in respect of the following items may be claimed (even though they may form an integral part of a building):

1   Electrical, cold water, gas and sewerage systems which are provided either to meet the requirements of a trade or to serve machinery or plant used for a trade.
2   Space heating systems, water heating systems, ventilation systems, air cooling and air purification systems and any ceiling or floor comprised in these systems.

3 Manufacturing and process equipment, storage equipment (including cold rooms), display equipment and counters, checkouts, etc.
4 Kitchen appliances, sanitary equipment and furniture and furnishings.
5 Lifts, hoists, escalators and moving walkways.
6 Sound insulation provided to meet the needs of a trade.
7 Computer installations, telecommunication systems and surveillance systems including cabling, etc.
8 Fire protection equipment and fire alarm systems.
9 Refrigeration and cooling equipment.
10 Any machinery.
11 Intruder alarm systems.
12 Movable partitioning which is intended to be moved in the carrying on of a trade.
13 Strongrooms (in banks and building societies) and safes.
14 Decorative items which are provided for customer enjoyment in hotels, restaurants and similar businesses.
15 Advertising hoardings, signs and displays, etc.

Although expenditure on fixed structures (works of a civil engineering nature) does not attract capital allowances, expenditure on the following assets may attract allowances under the classifications 'machinery' and 'plant':

1 The alteration of land for the sole purpose of installing plant and machinery.
2 The provision of dry docks.
3 Jetties and similar structures provided to carry machinery and plant.
4 Pipelines.
5 Towers to support floodlights.
6 Reservoirs incorporated in water treatment works.
7 Storage silos and tanks.
8 Slurry pits.
9 Swimming pools, including diving boards, slides, etc. and any supporting structures.
10 Fish tanks and ponds.
11 Rails, sleepers and ballast for a railway or a tramway.

The items that can be claimed, therefore, extend beyond those normally considered as plant by most surveyors. It should also be noted that consideration of available allowances should be taken into account at the design stage; for example, a plenum heating and cooling system may become a financially favourable alternative when the available capital allowances are taken into account in a comparative design study. However, as Collins[1] notes, 'there must . . . be no suggestion that the design is tax-led'. When considering the installation of machinery and plant in an existing building where a trade is carried on, the work incidental to this installation may be treated as machinery and plant.

Legal cases concerning the definition of machinery and plant can help in calculating and negotiating claims for allowances and in using tax planning as part of the design process to maximize the benefits available. Some of the decisions defining plant are summarized below:

1   In *Wimpey International Ltd v Warland 1988 CA*, it was decided that items of decoration in the company's restaurants were not classifiable as machinery and plant, as they were part of the setting.
2   In *Leeds Permanent Building Society 1982 CD*, it was held that decorative screens which incorporated the organization's logo were plant and machinery.
3   In *Hampton v Fortes Autogrill Ltd 1980 CD STC 80*, it was decided that a false ceiling installed as cladding for mechanical and electrical services installations was not plant and machinery.
4   Movable metal office partitioning was held to be plant in *Jarrold v Johngood & Sons Ltd 1962 CA 40 TC 681*.
5   A canopy covering the service area of a petrol station was held not to be plant in *Dixon v Fitch's Garage Ltd 1975 CD STC 480*.
6   In *Thomas v Reynolds etc 1987 CD STC 50*, it was held that an inflatable dome covering a tennis court was the setting in which business was carried on and was not plant.
7   A prefabricated building used by a school as a laboratory and gymnasium was held not to be plant in *St John's School v Ward 1974 CA 49 TC 524*.
8   In *Gray v Seymour's Garden Centre CHD 1993*, it was held that a special greenhouse was not plant.
9   A concrete silo with gantries, conveyors and chutes constructed by a grain importer was held to be plant in *Schofield v R & H Hall Ltd 1974 NI 49 TC 524*.

Capital allowances for machinery and plant are available on a reducing balance basis. The allowances are 25 per cent per annum. Generally, expenditure on machinery and plant is grouped together into a single pool; however, there are some exceptions (notably, short-life assets, cars, and assets for foreign leasing). If assets are sold or disposed of, the sale price or market value (assuming this is less than the original purchase price) is compared with the unrelieved expenditure. Where the sale price exceeds the unrelieved expenditure, the excess is added to taxable income as a balancing charge. A balancing allowance is available for the shortfall between sale price and unrelieved expenditure.

Example 1

A company has the following machinery and plant transactions in the accounting periods ending 31 December 1994 and 1995:

	£
April 1994 Purchase of plant	50 000
November 1994 Sale proceeds	5 000
February 1995 Purchase of plant	5 000
March 1995 Purchase of plant	110 000
June 1995 Sale proceeds	25 000
September 1995 Purchase of plant	16 000

The pool balance brought forward to 1 January 1994 is £55,000. The allowances are as follows:

For the accounting period ending 31 December 1994

	£
Pool value brought forward	55 000
Additions	50 000
	105 000
Less sales proceeds	(5 000)
	100 000
Writing down allowance @ 25%	(25 000)
Pool balance carried forward to 1995	75 000

For the accounting period ending 31 December 1995

	£
Pool value brought forward	75 000
Additions	131 000
	206 000
Less sales proceeds	(25 000)
	181 000
Writing down allowance @ 25%	(45 250)
Pool balance carried forward to 1996	135 750

Taxable profit for the accounting periods to 31 December 1994 and 1995 is therefore reduced by £25,000 and £45,250 respectively due to the available capital allowances for machinery and plant.

Different building uses will obviously attract different levels of allowance for machinery and plant. Figure 17.1 shows how capital allowances for machinery and plant could arise in an office.

Industrial buildings

Industrial buildings are defined as 'buildings used for the qualifying trades stated in the Capital Allowances Act 1990'. It is the use of the building that is important. Qualifying uses most commonly include buildings which are used for manufacturing and processing goods; mills, factories, etc.; buildings used for the storage of materials to be used in the manufacture of other goods; and buildings used for the repair and maintenance of goods (but generally not forming part of a retail business).

   Allowances for industrial buildings are presently 4 per cent per annum calculable on a straight-line basis from the time the building was first used as an industrial building. The allowances are available in respect of the cost of construction. This excludes the cost of purchasing the land, but does include the cost of site preparation. For the purposes of capital allowances, industrial buildings have a maximum life of twenty-five years; after this period of use as an industrial building no allowances are available. For buildings which were brought into use prior to 6 November

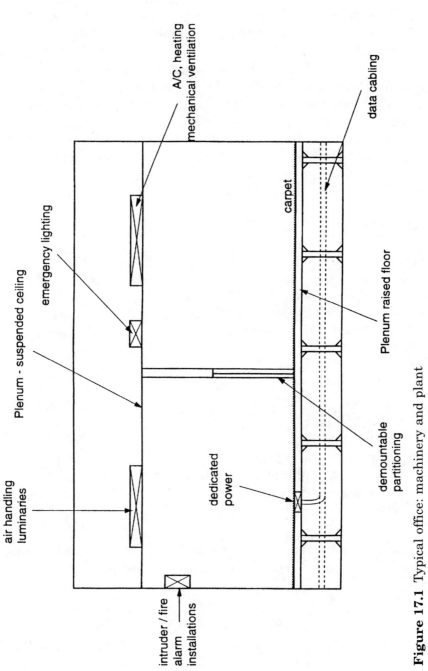

Figure 17.1 Typical office: machinery and plant

1962, the writing-down allowance is 2 per cent per annum and the tax life is fifty years. Initial allowances have, in the past, been available for the first year of use of a qualifying building.

If a qualifying building is sold during its tax life, a balancing adjustment is made in respect of allowances claimed by the vendor. Depending on the sale price, this could result in a *balancing allowance* or a *balancing charge*. The purchaser is entitled to allowances for the remainder of the building's tax life. The calculation of the relief available is, however, based upon the original construction cost of the building. Example 2 illustrates the position.

Example 2

In December 1988 the construction costs of an industrial building were £1,000,000, excluding associated land costs of £500,000. Annual allowances totalling £320,000 have been claimed for eight years' use. The building was sold in 1996 for £1,100,000.

The position with regard to industrial building allowances is as follows:

	£
Proceeds of sale	1 100 000
Less land cost	500 000
Cost of building	600 000
Original building cost	1 000 000
Cost of owning building	400 000
Allowances claimed	320 000
Vendor's balancing *allowance*	80 000

The remaining tax life of the building (ignoring months for simplicity) is 17 years. The purchaser would get the following annual relief:

$$\text{annual relief} = \frac{£600,000}{17 \text{ years}}$$

If the building had sold for £1,400,000 the position would be as follows:

	£
Proceeds of sale	1 400 000
Less land cost	500 000
Cost of building	900 000
Original building cost	1 000 000
Cost of owning building	100 000
Allowances claimed	320 000
Vendor's balancing *charge*	220 000

The remaining tax life of the building (ignoring months for simplicity) is 17 years. The purchaser would get the following annual relief:

$$\text{annual relief} = \frac{£900,000}{17 \text{ years}}$$

Hotels

Capital allowances are available against the construction cost of hotels. To qualify for relief, a hotel must be a permanent building which is open for at least four months between April and October, offer at least ten letting bedrooms and provide normal hotel services (i.e. breakfast, room cleaning, etc.). The writing-down allowance of 4 per cent per annum based upon initial construction cost is calculated on the same basis as that for industrial buildings. The situation in respect of balancing adjustments on the sale of a hotel building is also dealt with in a similar way to the adjustments for industrial buildings. In the period 1 November 1992 to 31 October 1993, an initial allowance of 20 per cent was available against qualifying expenditure.

Enterprise zones

Enterprise zones are areas so designated by the Secretary of State. Expenditure on commercial and industrial buildings within enterprise zones, but excluding expenditure on dwelling houses, is eligible for 100 per cent initial allowance (however, a lower initial allowance can be claimed). Expenditure either incurred or contracted for within ten years of the creation of an enterprise zone qualifies for relief. Buildings which are sold within their first twenty-five years are subject to the balancing adjustments described above for industrial buildings.

Financial impact for a development

A single development can attract more than one type of capital allowance. It is therefore important to plan any claim for relief to ensure that the maximum benefits will accrue. For example, allowances for a hotel development could be claimed on the basis of 4 per cent writing-down allowance on the whole construction cost. However, it may be more financially attractive to extract the qualifying expenditure on machinery and plant from the cost of the hotel and claim these separately on a 25 per cent reducing balance basis, thus receiving the relief sooner. The benefits of different tax planning strategies will be different for individual clients, and so tax planning should take account of a client's overall business status and objectives.

Value added tax

Value added tax (VAT) was introduced in the United Kingdom in the Chancellor's budget in 1972. VAT is a European tax which has resulted from membership of the European Union (EU) (formerly the European Economic Community (EEC)). The principal legislation is the Value Added Tax Act 1994 which, together with other statutory instruments, provides the framework for this tax. HM Customs and Excise are the enforcing body for VAT legislation. VAT is a tax on the supply of goods and services. There are three categories of supplies to consider:

(i) Standard-rated supplies

Standard-rated supplies are classified as supplies which are neither zero-rated nor exempt. Standard-rated supplies are subject to VAT at the rate of $17\frac{1}{2}$ per cent.

(ii) Zero-rated supplies

Those goods and services which are zero rated are defined in Schedule 8 of the Value Added Tax Act 1994. Supplies which are zero rated are subject to a positive rate of VAT of 0 per cent. There are sixteen classifications of zero-rated supplies, which can be summarized as follows:

(*a*) Food (with some exceptions).
(*b*) Sewerage services and water (except that for industrial use, which is standard rated).
(*c*) Books, newspapers, magazines, etc.
(*d*) Talking books for the blind and handicapped.
(*e*) Construction of new domestic buildings and approved alterations to domestic listed buildings.
(*f*) Protected buildings.
(*g*) Transport (except for taxis and hire cars, which are standard rated).
(*h*) Caravans and houseboats.
(*i*) International services.
(*j*) Bank notes.
(*k*) Drugs and medicines supplied on prescription.
(*l*) Children's clothing and footwear.
(*m*) Exports.
(*n*) Goods sold in tax-free shops.
(*o*) The transfer of gold between banks and similar organizations.
(*p*) The supply of donated goods by some charities.

(iii) Exempt supplies

Exempt supplies are not subject to VAT. They include:

(*a*) Land (with the exceptions noted later).
(*b*) Insurance.
(*c*) Postal services (by the Post Office).
(*d*) Betting, gaming and lotteries (but takings from game machines are standard rated).
(*e*) Financial services (with some exceptions).
(*f*) Education.

(*g*) Health and welfare services.
(*h*) Burial and cremation.
(*i*) Sports competitions.
(*j*) Fund-raising activities by charities.
(*k*) Works of art.
(*l*) Trade unions and professional bodies.

VAT is applicable to taxable supplies (all supplies, with the exception of exempt supplies) which are made in the course of a business by a taxable person. A *taxable person* is a person registered for VAT, and can include individuals, companies, clubs, etc. It is mandatory for an individual or organization to register for VAT if either:

(i)   at the end of a month the value of taxable supplies made in the previous twelve months exceeds £46,000 (unless the value of the next twelve months' supplies will not exceed £46,000); or
(ii) the value of taxable supplies to be made in the next thirty days is thought to be in excess of £46,000.

A business which is not legally required to register for VAT can apply to Customs and Excise for voluntary registration. This will be granted if the business makes taxable supplies. This could be financially beneficial in allowing a business to recover input tax and, providing its customers are VAT-registered, the addition of VAT on supplies should not affect trade. A person may apply for voluntary de-registration if taxable supplies are not expected to exceed £44,000 in the following twelve months.

The process

The intention of VAT is to make a charge on the final consumers of goods and services, and therefore businesses should not be affected. VAT charged by a business on its supplies is called *output tax*. The VAT paid by a business on purchases made for the manufacture of goods or the provision of services (i.e. on the factors of production) is called *input tax*. The system operates such that VAT-registered organizations can recover input tax, but must pay over output tax collected to Customs and Excise. Thus the difference between output tax and input tax for a business results in either a payment to Customs and Excise if output tax is greater than input tax, or a sum recoverable from Customs and Excise if input tax is greater than output tax. Generally, therefore, the burden falls upon the final consumer. Businesses are effectively responsible for the administration of the system and the collection of VAT.

Businesses that are not VAT-registered cannot recover input tax. Businesses that make a combination of taxable and exempt supplies can recover a proportion of their input tax. The significance of the three classifications of supplies, especially the different consequences of making zero-rated and exempt supplies, is now obvious. Although for both zero-rated and exempt supplies there is no output tax charge, if zero-rated supplies are made a registered person can recover input tax. A business making exempt supplies cannot recover input tax, and therefore the cost of input tax will be reflected in the cost of the supplies.

Example 3

At the end of an accounting period, a VAT registered company has made taxable supplies to a value of £2,500,000 for the period and has incurred taxable inputs for the period of £2,000,000. The VAT position is as follows:

	£
VAT on taxable supplies	
£2,500,000 @ 17½% (output tax)	437 500
VAT on taxable inputs	
£2,000,000 @ 17½%	350 000
Balance to Customs and Excise	87 500

Construction

With the exceptions noted above under zero-rated supplies, all construction activity, including new work, repair, maintenance, alteration, civil engineering and demolition, is classified as standard-rated supplies and therefore subject to VAT at 17½ per cent.

Land and buildings

The sale of land for domestic building is an exempt supply, whereas land for uses other than domestic building is classified as a standard-rated supply. The sale of used buildings, leases of used buildings, and leases for new domestic buildings are exempt supplies. The sale of new domestic buildings (where the vendor is the builder), and the sale of leases capable of exceeding 21 years for new domestic buildings, are zero rated. The sale of new non-domestic buildings is classified as a standard-rated supply.

References

1. COLLINS, C., Minimising the tax-man's share, *Estates Times*, 12 October 1990, p. 15.

Bibliography

1. HOMER, A. and BURROWS, R., *Tolley's Tax Guide 94–95* (Surrey: Tolley Publishing Co. Ltd, 1994).
2. KING, P. S. D., *VAT on Construction and Development Work* (London: Granada Publishing Ltd, 1984).
3. COLLINS, C., Minimising the tax-man's share, *Estates Times*, 12 October 1990, p. 15.
4. BLAKELEY, P., Capital allowances and the property owner, *Estates Gazette*, 29 September 1990, issue 9039, p. 26.
5. ROWE, R., Capital Allowances Planning to Maximise Tax Relief, *Property Journal*, June 1991, pp. 18–19.
6. ROWE, R., Capitalising on Allowances, *New Builder*, 23 August 1990, p. 20.

18

Collateral warranties

Introduction

Collateral warranties create contractual agreements between parties where they would otherwise not exist. The use of collateral warranties on construction projects has risen dramatically over the last decade. This increase may be attributed to a variety of factors, including:

1 Developments in the law of tort.
2 Changes in building procurement practice.
3 The ever increasing complexity of financial arrangements and investment interests associated with development projects.
4 The increase in multiple ownership of development projects.

As a result of this increase, it is now essential for surveyors to have a basic understanding of the law and practice relating to the use of collateral warranties. Surveyors may be required to provide collateral warranties in respect of the services they are offering to the benefit of third parties, and so a clear understanding of the implications of such agreements is essential. Surveyors may be required to give advice to their clients with regard to obtaining collateral warranties for a development project, so, again, an understanding of collateral warranties is necessary. As Pike[1] notes, 'it is incumbent upon design professionals and other advisers of property investors to ensure that appropriate collateral warranties are obtained. If they do not, they may be held liable to their clients for negligence'.

This chapter defines collateral warranties, explains why they may be required, considers who may require them and from whom they may be required, and discusses some of the common provisions contained in them.

General principles

Consider the situation where a leaseholder has just leased a building from a developer on the basis of a full repairing and insuring lease. On discovering design defects, what action can the leaseholder take? The leaseholder, while being responsible to the developer for rectifying the defects, would want to recover the cost from the negligent architect. But, in the absence of a collateral warranty, there is no contractual relationship between the leaseholder and the architect. Until recently, it was assumed that there existed a remedy for the leaseholder in tort. However, recent cases, in particular the decision in *Murphy v Brentwood District Council [1990] 2 All ER 908,* have restricted liability in tort and created some uncertainty about rights in tort. The response to this has been an increase in demands for collateral warranties from all those who may have an interest in a development, thus providing a contractual remedy in such situations.

Consider the situation of a contractor's insolvency during the course of a project. Does the Employer have the right to the benefits of sub-contract agreements to enable completion of the project with minimum disruption? In the absence of contractual relationships with the sub-contractors (which could be provided by collateral warranties), the answer is No.

Collateral warranties create a direct contractual link between parties where one would otherwise not exist. They are agreements alongside (collateral to) another contract. So, typically, for a surveyor, a collateral warranty will be collateral to the terms of engagement. For a contractor, a collateral warranty will be collateral to the building contract (and design supplement, where applicable). For a sub-contractor, a collateral warranty will be collateral to the sub-contract agreement (and design supplement where applicable). Collateral warranties give contractual remedies to parties who, as a consequence of the doctrine of privity of contract, would not otherwise have them.

The presentation of collateral warranties can vary from standard forms of agreement to simple letters from the warrantor to the warrantee. The names used to describe these agreements also vary; collateral warranties are sometimes referred to as duty of care deeds, deeds of responsibility, etc. However, for a collateral warranty to exist, the normal prerequisites for a contract (offer and acceptance, consideration and the intention to create legal relations) must be observed.

Warranties can be created by either:

(i)   a simple contract requiring consideration; or
(ii) as a deed, in which case consideration is not required, but there are some formalities to be performed with regard to the execution of the deed.

In the absence of specific limitation wording, the effective difference between a simple contract and a deed is the limitation period. With a simple contract, the time limit for pursuing a claim is six years from the date of breach. For a deed, the period is twelve years. Thus it would be inappropriate to have a collateral warranty created by deed running alongside a main agreement with a limitation period of six years, as the warranty would be more onerous than the main agreement.

It is the view of many legal experts that the liabilities created by collateral warranties are greater than those that exist in tort. For example, designers who demonstrate reasonable skill and care would not be considered negligent in tort, whereas with a collateral warranty the standards required may, depending on the specific wording used in the warranty, be higher than this. In tort, with the exception of reliance duties, economic losses (for example, trading losses) are not recoverable, but under a collateral warranty, in the absence of an express limitation provision, economic losses are recoverable; the liabilities are therefore potentially greater.

Interested parties

Any individual or organization with a financial interest in a development project, but without a direct contractual relationship with the designers or builders, may seek the protection afforded by collateral warranties from these parties.

Consultants, contractors, sub-contractors and suppliers all have the potential to cause loss by their errors. Under certain circumstances, all may be required to provide a warranty to the benefit of a third party. The need for, and the format of, warranties will vary, depending on the nature of the project, the contractual relationships and the procurement strategy adopted.

Under traditional forms of procurement, contractors are responsible for the standard and quality of workmanship and materials and for the completion of work within a prescribed time period. This responsibility extends to include the work of sub-contractors and materials from suppliers. Collateral warranties may be required from contractors by funding institutions (who have an interest in the successful completion of the project and may want the right to act as Employer in, for example, the event of developer insolvency), by future tenants (particularly those tenants with a full repairing and insuring lease) and by future purchasers. Where a sub-contractor or supplier is required to undertake a design role, however, the contractor usually has no responsibility for the design undertaken by the sub-contractor or supplier. Because the agreement in such instances will be between the sub-contractor or supplier and the main contractor it would be advisable for the Employer (and, indeed, any other interested party) to obtain a collateral warranty from the sub-contractor or supplier in respect of the design work.

Consultants generally have responsibilities for design and management. For a typical commercial development, consultants will be appointed by a developer with whom they will be in direct contract. Again, warranties may be required by funding institutions, future tenants and future purchasers of the development.

Figure 18.1 shows the typical contractual arrangements for a development project procured using the traditional approach. Some of the possible situations where collateral warranties may be used or required are also shown in Figure 18.1. If there are many tenants or purchasers, or if there are many funders with an interest in the project, then a large number of separate collateral warranties may be required; managing warranties could then become complex (and expensive).

Under other forms of procurement, the situation regarding warranties will vary. When advising clients on the need for protection by way of warranties, surveyors should adopt a

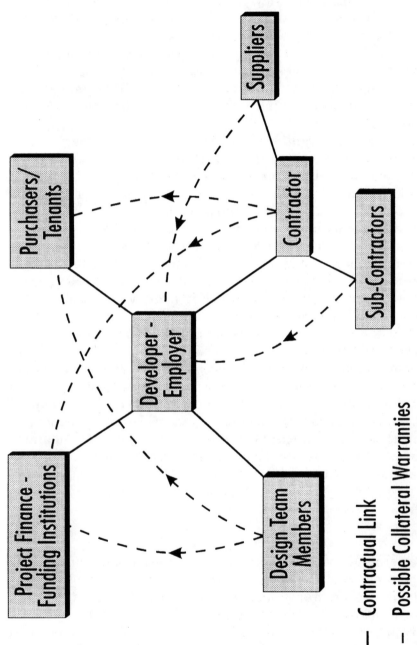

Figure 18.1 Contractual arrangements for traditional procurement

—— Contractual Link

– – – Possible Collateral Warranties

systematic approach. They should identify the likely causes of loss and who might be responsible for the loss arising. They should then examine the contractual arrangements to identify the protection afforded without collateral warranties, and where insufficient protection is available they should consider using them. As noted by Kitchen[2], 'a collateral warranty is simply a way to shift responsibility for certain losses from one party to another'.

If collateral warranties will be required, it should be so stated in the main agreement. For example, if consultants are required to provide warranties for funders, tenants and purchasers, it should be stated in their terms of engagement. The information included in the main agreement should be such that the wording of the warranties and the beneficiaries are stated. This will enable the consultants to include in their fees for the costs, in terms of additional administration, professional indemnity insurance costs, etc., of providing the warranties. If insufficient, or no, provision for providing warranties is made within the original terms of engagement, then consultants are at liberty to demand additional payment or to refuse to provide warranties.

There are, therefore, implications for surveyors involved with drafting tender and contract documentation. If warranties are required from contractors, sub-contractors or suppliers, it should be established prior to tender. The number and format of warranties required should be clearly identified in the tender and contract documentation. If this is not observed, then obtaining warranties may prove to be either prohibitively costly or difficult (they may even be refused).

Terms of warranty

The terms of warranty depend upon the specific nature of a project and the benefits required. Generally, the obligations imposed by a warranty should be the same as those in the main agreement; it is normal to refer to the main agreement in a warranty. Therefore, if an architect is in breach of contract to a developer, he could also be in breach of warranty with a tenant. The obligations imposed by a warranty should be no greater than those imposed by the main agreement. Also, care should be taken to ensure that there are no terms included in warranties that are in conflict with terms in the main agreement. Although the format and composition of warranties varies enormously, the following are areas commonly dealt with:

1 *Limitation of damages* – in view of the potentially greater liability in contract than tort, it is common to restrict the damages recoverable, e.g. covering the cost of remedying defects only.
2 *Use of reasonable skill and care in constructing or designing.*
3 *Design or construction fit for its intended purpose.*
4 *Insurance* – for example, consultants may be required to take out and maintain professional indemnity insurance. Warrantors should ensure that they inform their insurance companies with regard to all collateral warranties they provide to ensure that their insurance cover is valid.
5 *Assignment* – occurs where the rights or obligations of one of the parties to a collateral warranty are transferred to a third party, for example, a second tenant. The law associated with the assignment of warranties without consent is complex. In this respect, it is appropriate to make express provision within a warranty with regard to assignment; it should state whether the warranty can be assigned and, if so, how, to whom and how many times.
6 *Limitation* - to ensure that a warranty is no more onerous than the main agreement to which it relates, it is sometimes advisable to include a clause limiting the validity of the warranty from a fixed date, for example, the date of the main agreement, for a specified period of time.
7 *Novation rights* – for example, the right of an Employer to take over a sub-contractor's appointment in the event of a breach of contract or sub-contract by the main contractor.
8 *Preservation of rights at common law* – a specific clause is common.

In situations where joint liability could occur (for example, bad workmanship by a contractor coupled with negligence on the part of the contract administrator in the supervision of the Works), warrantors should ensure that warranties are provided by all those with whom they may share joint liability. If this is not ensured, the total liability to the warrantee will be borne by the single warrantor.

Any person or organization that is either bound by a warranty or entitled to enforce a warranty must be a party to that warranty. Thus, with a warranty for architectural design to the benefit of a funder, the parties to the warranty will be the architect (the warrantor), the funder (the warrantee), and the developer (as the funder will want the right to take over the architect's appointment in the event of breach by the developer, the developer will need to be a party to the warranty to grant this right).

There is a large, and increasing, number of standard forms of warranty available for use on construction projects. Many of these have been drafted by trade, professional and client organizations (for example, the Royal Institute of British Architects, the British Property Federation, etc.) to protect the interests of their members. Additionally, the use of bespoke warranties drafted by solicitors to cater for the interests of their clients is widespread. In all situations, whether responding to a request to provide a warranty or advising on the use of warranties, the obligations imposed by warranties should be carefully considered to ensure that they do not impose unrealistic obligations on the warrantor or fail to provide the benefits required of them.

An alternative to collateral warranties

For problems occurring after practical completion of a project, a form of insurance cover, often referred to as Building Users' Insurance against Latent Defects (BUILD), is becoming popular with developers. As Paterson[3] notes, this insurance 'is a non-cancellable material damage policy for the benefit of the developer, subsequent owners and occupiers'. The insurance deals with many of the concerns of tenants and purchasers for which warranties are used, but as it is only effective from practical completion, it does not deal with all the relevant issues.

From a contractor's or consultant's point of view, there may still remain some liability in respect of the insurers' rights of subrogation with BUILD unless either the contractor and consultant are named parties in the insurance policy or subrogation rights are waived by the insurers.

References

1. PIKE, A., Collateral Warranties under construction and civil engineering contracts, *Architect and Surveyor*, March 1990, p. 19.
2. KITCHEN, S., Paper Mate, *Building*, 31 January 1992, p. 20.
3. PATERSON, F. A., *Collateral Warranties Explained* (London: RIBA Publications Ltd, 1991).

Bibliography

1. PATERSON, F. A., *Collateral Warranties Explained* (London: RIBA Publications Ltd, 1991).
2. WINWARD, Fearon, *Collateral Warranties – A Practical Guide for the Construction Industry* (London: BSP, 1990).
3. KITCHEN, S., Paper Mate, *Building*, 31 January 1992, pp. 20–21.
4. CHAPPELL, D., Collateral Warranties – RIBA and RIAS Forms, *Architects Journal*, 26 July 1989, pp. 67–9.
5. PIKE, A., Collateral Warranties under construction and civil engineering contracts, *Architect and Surveyor*, March 1990, pp. 19–21.
6. CLARKE, R., Collateral Warranties – the current state, *Chartered Quantity Surveyor*, April 1990, pp. 9–11.

19

Guarantees and bonds

Introduction

The costs to an Employer following the failure of a contractor during the course of a project can be significant. Similarly, in today's industry, a contractor's success is often dependent on the performance of sub-contractors. The failure of a sub-contractor can mean a potentially profit-making contract resulting in loss for a contractor. The objectives of guarantees and bonds are to encourage performance and to provide financial recompense in the event of failure.

This chapter examines the purpose of guarantees and bonds, discusses their various sources and formats, considers the arguments for and against their use, and discusses situations where their use may be appropriate.

The purpose of guarantees and bonds

Guarantees and bonds are arrangements whereby the obligations that one party owes to another under an agreement are guaranteed by a third party. The third party, usually an insurance company or a bank, promises to pay a sum of money in the event of non-performance. The following types of bonds and guarantees are often used on construction projects:

Performance bonds/guarantees

These are, as the name implies, aimed at ensuring the satisfactory performance of a party to a contract. They are the most commonly used type of bond/guarantee used in the construction industry. They may be required by an Employer to ensure the performance of a contractor, or by a contractor to ensure the performance of a sub-contractor. In the event of non-performance, the guarantor or bondsman provides financial compensation. The usual limit of compensation is 10 per cent of the contract sum.

Tender bonds/guarantees

These are obtained to ensure that a successful tenderer will enter into a contract. Their value varies, but is usually somewhere between 1 per cent and 5 per cent of the tender sum. One of the purposes of tender bonds/guarantees is to provide financial recompense for the costs associated with the refusal by a tenderer to honour a tender. Most typically, they are requested by a contractor to ensure a sub-contractor's commitment to enter into a sub-contract agreement. The associated costs and disruption to a contractor's programme in the event of a sub-contractor withdrawing an offer can be significant. A new sub-contractor will have to be found and appointed, which may cause delays to the programme resulting in liquidated damages becoming payable to an Employer and/or acceleration costs for the completion of the project. It may also not be possible to obtain such a competitive tender the second time around (and, probably, at short notice), and thus the project costs will increase, leading to a reduction in profit. A tender bond/guarantee may discourage withdrawal by a sub-contractor, or at least provide some financial recompense to a contractor in the event of withdrawal.

Payment bonds/guarantees

These ensure the obligation to make payment. They may be requested by contractors to guarantee an Employer's duty to make payments or by sub-contractors to ensure payment from a contractor.

In most situations, the protection afforded by guarantees and bonds will be required in the event of insolvency. The effects of insolvency on construction projects are discussed in detail in Chapter 20.

Sources and formats of bonds

The two most commonly used sources of protection are bank
guarantees and surety bonds. The terms 'guarantee' and 'bond'
are used synonymously within the industry. There is, however, a
difference in the definition of the two terms. McDevitt[1] draws the
distinction between demand guarantees and performance bonds.
To summarize:

Demand guarantees are documentary covenants made by a
guarantor to indemnify a beneficiary, subject to the conditions in
the covenant. The guarantee is an agreement between the
guarantor and the beneficiary. Thus, if an Employer is given a
demand guarantee by a bank in respect of the obligations of a
contractor, the contractor is not a party to the guarantee
agreement. The beneficiary is therefore in a strong position
should there be a default. Demand guarantees are contracts and
can be created by either a simple contract or executed as a
deed.

Demand guarantees are usually given by banks. There are two
basic types: on demand guarantees (often referred to as on
demand bonds) and documentary demand guarantees. On
demand guarantees basically require a guarantor to make
payment to a beneficiary upon request to do so. In the case of
documentary demand guarantees, payment will only be made on
the furnishing, by the beneficiary, of the documents required by
the terms of the guarantee. These, for example, may be documents proving a court judgment.

Banks favour demand guarantees because they do not need to
get involved in legal arguments and disputes following a
default; their position is generally straightforward. However,
the situation is not so satisfactory for those required to provide
demand guarantees. Take, for example, a contractor required to
provide a 10 per cent demand guarantee in respect of a
£2,000,000 contract. The guarantee will be for the sum of
£200,000. The contractor's bank issuing the guarantee will treat
the value of the guarantee as contractor's credit and will,
therefore, reduce any credit facilities offered to the contractor by
this sum. In addition, the bank will probably require security
from the contractor to support the credit. Both these actions will
affect a contractor's cash flow and make it more difficult for him

to perform contracts. Indeed, the operating capacity of a construction firm can be reduced by the requirement to provide demand guarantees. A contractor in this position may also feel exposed, especially where on demand guarantees are provided. The contractor has minimal rights to prevent a bank paying against an on demand guarantee. Banks will pay on demand and leave the contractor to settle any dispute directly with the beneficiary.

Performance bonds are three-party agreements between a surety (or bondsman or guarantor), a beneficiary and a principal debtor. In essence, the surety guarantees the performance of the principal debtor. In the event of a default by the principal debtor, the surety will make good the loss caused to the beneficiary up to the financial limit of the bond. Performance bonds will usually be provided jointly and severally in the name of the principal debtor and the surety; the surety would, therefore, rely first on the principal debtor satisfying any claim before becoming involved. The principal debtor receives no protection from the surety under the bond; it is not an insurance policy. Thus, the only situation where a surety is likely to be required to make payment would be in the event of the insolvency of the principal debtor.

Performance bonds are usually issued by insurance or surety companies. Unlike banks, these organizations will have little information regarding the financial standing of a principal debtor. Surety companies will, therefore, usually undertake extensive checks to ascertain the risk associated with providing a bond. The cost of the bond, together with the knowledge that an organization is able to obtain a bond, will often provide some comfort to a beneficiary that a principal debtor is sound.

Unlike demand guarantees, the beneficiary of a performance bond will be required by the surety to prove loss before payment is made. Additionally, the bond will not affect a principal debtor's credit facility, nor will it usually be necessary to provide security. As with demand guarantees, bonds can be created by simple contract or deed.

Performance bonds are often required by Employers in respect of a contractor's obligations under a construction contract. Similar protection may be required by contractors from their sub-contractors.

The arguments for and against bonds and guarantees

If good practice is observed in the selection and appointment of contractors and sub-contractors, it could be argued that there is no need to obtain bonds and guarantees. Certainly, the requirement to provide protection in the form of guarantees or bonds will, in most cases, add to the cost of a project. Where a demand guarantee is required from a contractor's bank, the resulting cash flow problems have been discussed above.

In considering the need for bonds, a sensible approach is required. While the comfort provided to Employers by a guarantee or bond may be reassuring, the costs of providing the protection may not be worth the benefits obtained. Consider the situation where an Employer is a frequent developer or has a substantial development programme; the cost of obtaining performance bonds or guarantees for all contracts would probably be high compared to the value of claims resulting from those bonds. It may be more appropriate, therefore, for the Employer in this situation to accept the risk of contractor default rather than require a large number of expensive bonds.

The use of bonds

There is no definitive list of situations where a bond should be used, and a decision will need to be made for each project. In some circumstances, a client may insist that a contractor provides a performance bond; in others, the design team may need to advise a client on the availability, suitability and cost of the various forms of protection available. Situations where a bond may be particularly appropriate would include:

(i)   to protect the interests of a 'one-off' developer;
(ii) where a new or unproved contractor is involved in a project;
(iii) tender bonds in respect of nominated sub-contracts;
(iv) where a bond is thought appropriate for the risks inherent in a project.

If a bond will be required, it should be clearly stated in any tender and contract documentation. The terms of the bond should include the financial limit, details regarding the release of the bond, and defaults and non-performance covered by it. Care should be taken to ensure that the conditions of the bond are observed; for example, giving notices of default to the surety. Failure to do this may render a claim invalid. The alteration of contract terms will render a bond invalid (in the absence of express conditions in the bond permitting alteration).

Where a surveyor is requested to check the wording or form of a non-standard bond, specialist advice should be sought. Standard forms of bond available include those published by the ICE, FIDIC and ABI. The use of standard forms of bond is recommended. There is no JCT standard form of performance bond.

References

1. MCDEVITT, K., *Contract bonds and guarantees*, The Chartered Institute of Building Occasional Paper No. 34 (1985), Berkshire.

Bibliography

1. MCDEVITT, K., *Contract bonds and guarantees*, The Chartered Institute of Building Occasional Paper No. 34 (1985), Berkshire.
2. POOLE, A., A clear view of the pitfalls, *Building*, 17 November 1995, p. 40.
3. RIDEOUT, G., Licence to kill, *Building*, 5 March 1993, pp. 20–21.
4. MINOGUE, A., Gentlemen don't prefer bonds, *Building*, 9 September 1994, p. 38.
5. SPENCER, K., No need to pay on demand, *Building*, 28 July 1995, p. 39.
6. NJCC, *Guidance Note 2 – Performance Bonds* (London: RIBA Publications Ltd, 1986).
7. Association of British Insurers, *ABI Model Form of Guarantee Bond – An Explanatory Guide*, September 1995.

20

Insolvency

Introduction

Unfortunately, insolvency is a common occurrence within the UK construction industry. Its causes are various. Certainly, the ease with which contracting businesses can be created, due in part to the minimal amount of capital investment required to start a construction firm, often gives rise to some very fragile arrangements. The capital goods nature of construction industry products, together with the large proportion of industry output flowing from often widely fluctuating and unpredictable levels of public expenditure, has created a 'boom and bust' construction industry cycle. This is often cited, particularly at times of recession, as a cause of mass insolvency.

Competitive tendering practices used throughout the industry have also been cited to explain the high incidence of contractor insolvency. The quality of management expertise within the industry is generally considered poor in comparison with most other industries. There are, certainly, many cases of contractor insolvency which have resulted from the mismanagement of a business as opposed to unfavourable external conditions. Whatever the causes of insolvency, however, the situation is unlikely to change in the short term. Surveyors must, therefore, be equipped to deal with the incidence of contractor and sub-contractor insolvency.

The damage resulting from insolvency can be widespread. The cost, quality and duration of a construction project can all be detrimentally affected as a consequence. Additionally, looking down the contractual hierarchy, the solvency of sub-contractors and suppliers can also be damaged following the insolvency of a contractor; thus one insolvency can affect many projects. The failure of one large British contracting organization during the recession of the 1980s resulted in a long chain of insolvencies of associated sub-contractors and suppliers. The insolvency of a developer can create financial difficulties for the contractors and, consequently, the sub-contractors and suppliers involved with his developments. Again, these problems can manifest themselves on other projects for other clients.

It is important for surveyors to ensure, as far as they can, that the contractors and sub-contractors they appoint are financially stable and are likely to remain so for the duration of their contracts. Obviously, this can never be guaranteed, but it is prudent to make formal and informal checks prior to contract or, preferably, tender. In the unfortunate event of contractor insolvency during the course of a project, surveyors must advise and act to protect the interests of their clients. This will mean ensuring that projects are completed, where appropriate, with minimum loss of value to their clients in terms of increased time, additional cost and reduced quality. (See section on *Methods of Contractor Selection*, p. 45.)

This chapter provides a brief introduction to the law of insolvency. It examines the provisions in JCT 80 insofar as they apply to the insolvency of either the contractor or the Employer. It considers the action and procedures to be adopted leading up to, and following, contractor insolvency. Finally, it discusses the situation of insolvency and nominated sub-contractors.

The law

The main statute covering insolvency law is the Insolvency Act 1986 (the Act). This legislation was introduced to implement some of the recommendations of the Cork Report. The Act has improved insolvency practice by controlling those persons entitled to act as insolvency practitioners and by introducing two new procedures, the administration order and the voluntary arrangement, which, as Newman[1] notes, 'aim to preserve a business in some form or other. This is a major departure from the emphasis of traditional insolvency procedures on extinguishing businesses'. There are, of course, many other statutory instruments and the like, together with much case law, pertinent to insolvency. It is not within the scope of this chapter or, indeed, a book of this nature to identify and explain the law in detail. Nor is it essential for the majority of surveyors to have this level of detailed knowledge. However, this section will provide a general introduction to the legal meaning of insolvency which, in itself, is often a source of confusion.

As the majority of contracting organizations are companies of one form or another, this chapter deals with the law relating to corporate insolvency. It should be noted that the law and procedures are different for individuals and partnerships.

Insolvency is broadly concerned with the inability to pay debts. There are various situations which can arise under the Act.

(i) Liquidation

This refers to the winding-up of a company. Trading usually ceases upon liquidation; the assets of the company are collected and used to offset liabilities. There are two categories of liquidation – *voluntary liquidation* and *compulsory liquidation*.

Voluntary liquidation can be either members' or creditors' voluntary liquidation.

Compulsory liquidation is the result of a court order for a company to be wound up. This is usually because the company is unable to pay its debts. There are two tests that can be used to determine if a company is unable to pay its debts. First, the going concern or cash flow test, whereby a company is judged to be insolvent if it cannot pay debts as they become due. Secondly, the balance sheet test, which is a more long-term test. The balance sheet test examines the value of a company's assets in relation to the amount of its liabilities. If assets are less than liabilities, then the company will be deemed insolvent. The nature of construction – in particular the difficulties of accounting for work in progress – can make the application of this second test difficult.

A liquidator will be appointed to take control of the winding-up of the business. For compulsory liquidation this will, in most cases, be the Official Receiver. The proceeds of liquidation are distributed in accordance with the following hierarchy:

(*a*) fixed charge holders;
(*b*) liquidators' fees and expenses;

(c) preferential creditors – Inland Revenue; Customs and Excise; National Insurance contributions; pension schemes; employees' pay; etc.;

(d) floating charge holders;

(e) unsecured creditors;

(f) shareholders.

Once a winding-up petition has been presented by creditors, they can apply to a court for a provisional liquidator to be appointed.

(ii) Administrative receivership

An administrative receiver can be appointed by a debenture holder (usually a bank) in accordance with a provision in the debenture agreement regarding a specified situation occurring. A debenture is a security given to a lender against borrowings. The duty of the administrative receiver is to recover the borrowings owed to the debenture holder by realizing those assets which are the subject of the debenture. In some circumstances, it may be advantageous to the debenture holder for the administrative receiver to keep a company operational if more of a debt may thus be realized.

(iii) An administration order

This will be made on the successful petition by a company, its directors or creditors for an order. An administration order is aimed at rescuing rather than winding up a company. A court will issue an administration order if it is satisfied that a company is either unable, or likely to become unable, to pay its debts, but only if a benefit is likely to arise as a consequence of the order (for example, the survival of the company or a better financial position than would be achieved with liquidation). An administration order freezes the affairs of a company, thus allowing breathing space for corporate rescue to be investigated.

Once an order has been made, a company will be managed by an administrator who will discharge the duties of the directors. It is the administrator's duty to manage the company in the interest of the creditors. Administrators have wide-ranging powers, and are generally permitted to do anything necessary to manage the business. Within three months of the date of an administration order, the administrator must submit proposals to creditors with regard to achieving the purpose of the administration order.

(iv) Voluntary arrangements

These enable the directors to make a proposal to their company and its creditors with regard to the payment of debts. The creditors are invited to a meeting at which a decision is made with regard to the acceptability of the proposed arrangements.

Provisions in JCT 80

Clauses 27 and 28 of JCT 80 contain the provisions dealing with either the insolvency of the contractor or the Employer. Amendment 11, issued in 1992, brought JCT 80 into line with the Act and with insolvency practice in general. Prior to Amendment 11, a contractor's employment was automatically determined in the event of insolvency. JCT 80 (as amended) now prescribes, in the event of contractor or Employer insolvency, either automatic determination of, or the right to determine, the contractor's employment. Whether determination is by choice or is automatic is dependent upon the classification of insolvency.

Contractor insolvency – automatic determination (clause 27.3.3)

The employment of the contractor is automatically determined if a winding-up order is made, or a provisional liquidator is appointed, or a trustee in bankruptcy is appointed, or a resolution is passed for voluntary winding-up (except where this is for the purposes of either reconstruction or amalgamation). The contractor (through the insolvency practitioner) and the Employer can, however, agree to reinstate the contractor's employment.

Contractor insolvency – the Employer's right to determine (clause 27.3.4)

If the contractor has an administrator or an administrative receiver appointed, the Employer is given the right to determine the employment of the contractor by issuing notice. The Employer may wish in such circumstances to appoint a new contractor to complete the Works; in this event, the right to determine is exercised. This decision may depend upon the specific require-ments of the Employer, the stage of the project, and the likelihood of suitable alternatives being offered by the receivers. However, if the Employer reserves this right of determination, there may be more scope for the receivers to sell the contractor's business, including contracts in progress, to another contractor, which might be an advantageous course of action for an Employer. Care should be taken, in exercising the right to determine, to ensure that an Employer's losses are not increased above what may be offered by the receivers.

In the event of receivers being appointed, the Employer is not required to make any further payments until either:

(i)   the Employer has exercised the right to determine the employment of the contractor, or
(ii) an agreement (known as a (clause) '27.5.2.1 agreement') has been made between the Employer and the contractor in respect of the continuation, novation or conditional novation of the contract. In this case the agreement will detail the arrangements for payment and the Employer's right to determination under clause 27.3.4 ceases.

Clause 27.5.3, however, permits the Employer and the con-tractor to make an interim arrangement for work to be carried out prior to one of the above events occurring. The Employer is also empowered to take reasonable measures to protect the Works and site materials during this period, the cost of such measures being deductible from sums due to the contractor.

The consequences of determination by the Employer

Clause 27.6 details the contractual consequences of determination by the Employer. The Employer is granted the right to employ other persons to complete the Works and to make good defects. Use of materials, plant, etc., on site is permitted with the proviso that, where they are not owned by the contractor, the consent of the owner is obtained. The requirement for the contractor to assign the benefits of supply and sub-contract agreements to the Employer upon determination does not apply in respect of determination flowing from contractor insolvency (clause 27.6.2.1). There is a similar restriction, in the event of a contractor's liquidation, on the Employer's right to make direct payments to suppliers and sub-contractors in respect of materials delivered or work undertaken prior to determination. The contractor must remove any temporary buildings, plant, tools, materials, etc., which belong to him and make arrangements to have items not owned by him removed if so requested by the architect.

It is common for materials to be supplied to a contractor on the condition that ownership of the goods does not pass to the contractor until the goods have been paid for. A retention of title claim (or Romalpa) will fail if the materials have been fixed in place. Further protection from retention of title claims is provided by JCT 80 in that the transfer of title in unfixed materials and goods passes to the Employer when the Employer has paid for them in interim payments. However, the situation is not so clear in respect of sub-contractors' unfixed materials.

The financial settlement, assuming the Employer completes the work, is calculated in accordance with the provisions in clauses 27.6.4 and 27.6.5. Generally, the Employer is not required to make further payment to the contractor (with some noted exceptions) until the Works are complete, defects are made good, and the account has been prepared. The account should detail:

(i)   the total expenses to the Employer following determination (including any direct loss or damage caused as a result of the determination);
(ii) payments made to the contractor; and
(iii) the amount that would have been payable had the Works been completed under the contract (i.e. a notional final account).

If the sum of (i) and (ii) is greater than (iii), the difference is a debt payable by the contractor to the Employer. If less, the difference is a debt owed to the contractor by the Employer.

Clause 27.7 deals with financial settlement where the Employer elects not to complete the Works.

Employer insolvency

In the event of Employer insolvency, a contractor is empowered, under clause 28.3, to determine his employment under the contract. The consequences of determination are set out in clause 28.4; they require:

(*a*) the contractor to remove (and ensure that sub-contractors remove) temporary buildings, plant, materials, etc., from the site;

(*b*) the Employer to pay to the contractor the Retention deducted from payments made to the contractor within 28 days of determination (subject to any right of deduction the Employer may have had prior to determination);

(*c*) the contractor to prepare an account detailing:
- (i) the total value of work completed;
- (ii) sums in respect of direct loss and/or expense payable under clauses 26 and 34.3;
- (iii) the cost of removing temporary buildings, plant, materials, etc., from the site;
- (iv) direct loss and damage caused by the determination;
- (v) the cost of materials and goods ordered and either paid for by the contractor or which the contractor is legally bound to pay for; ownership of these materials will pass to the Employer on payment by the Employer;

(*d*) the difference between the account and the amounts paid to the contractor is to be paid by the Employer within 28 days of presentation of account (no Retention is to be deducted). In practice, determination following insolvency may make this payment unlikely. In such circumstances the contractor will become an unsecured creditor of the Employer.

An Employer's interest in Retention is as a trustee for the contractor. The private versions of JCT 80 require the Employer to place retention monies in a separate bank account (designated to identify the amount as Retention held by the Employer on trust) if so requested by the contractor or any nominated sub-contractor. In the event of Employer insolvency, the separate bank account will ensure the trust status of Retention and should therefore protect a contractor's right to the Retention. In the absence of a separate account, case law suggests that a contractor may lose this right. It is, therefore, strongly recommended that, as a matter of course for all contracts, contractors (and sub-contractors) should request that Retention be held in a separate identified bank account by the Employer.

Avoidance, precautions and indications

Avoiding the incidence of contractor insolvency is obviously to be preferred. Where selective tendering is the basis for a contract, the pre-selection process should endeavour to check the financial status of those contractors under consideration. With open tendering, the checks must be made during the tender appraisal stage. There are no hard and fast rules regarding the checks to be made. The *Code of Estimating Practice*[2] gives guidance on pre-selection good practice. The following are common sources of information which can help to assess a contractor's financial standing:

(i)   information that members of the design team may have regarding individual contractors;
(ii)   information from financial checking agencies;
(iii) published accounts;
(iv)   bank references;
(v)   other references.

Precautions can be instigated to soften the blow of contractor insolvency. These may include the use of performance bonds and guarantees (as discussed in Chapter 19) and contractual Retention funds (most standard forms of construction contract prescribe retention of a proportion of interim payments pending satisfactory completion).

The signs of impending insolvency often become apparent some time in advance. If the signs are spotted then, at best the situation may be avoided, or, at least the impact of the insolvency may be reduced by taking pre-emptive action. In the run up to, and during the course of, a contract, surveyors should be aware of any signs that may signal trouble. These may include:

(i)   industry rumours;
(ii) slowing down of progress on site;
(iii) complaints of non-payment from nominated sub-contractors;
(iv) reduction in personnel on site;
(v)   reduction in amount of materials on site;
(vi) over-ambitious requests from a contractor for payment.

Surveyors must be careful in acting upon these signs. Impending insolvency can be accelerated, or sometimes caused, by the spreading of rumours suggesting financial instability. Once news gets around, creditors soon start calling in debts, and thus a contractor's difficulties can be intensified. So, if there are concerns, they should be kept within the design and client team until such time as the situation becomes official.

Action in the event of contractor insolvency

Action on suspicion of contractor insolvency

If the design team suspects that a contractor is suffering financial difficulties, there are some measures that should be considered to safeguard the Employer's interests should these suspected difficulties result in insolvency. Interim valuations should be carefully prepared to ensure that work is not over-valued. Materials on site should be carefully checked to ensure that they are in accordance with the contract, properly stored, intended for incorporation into the Works and that it is not unreasonable for them to be on site in relation to the programme. Checks should be made that nominated sub-contractors have been paid amounts stated as due in interim certificates. Suspicion does not give the right to undervalue work, but it would be prudent for surveyors to pay more attention to ensuring the accuracy of their interim valuations in such circumstances.

Immediate action upon contractor insolvency

It is important to act quickly following the announcement of contractor insolvency. News of insolvency can travel very quickly, and creditors, in an attempt to mitigate their losses, may try to recover debts by removing materials and equipment from sites. The following measures should be considered by the design team:

1   Advising clients of contractor insolvency, the contractual position, recommended action and liability.
2   Securing the site and considering employing a security firm to provide 24-hour security, changing the locks, carefully securing all valuable equipment, goods and materials.
3   Preparing a detailed valuation of completed work and an inventory of materials and equipment on site.
4   Stopping the processing of any payments to the contractor.
5   Contacting key sub-contractors and suppliers and commencing discussions concerning continuation contracts.
6   Checking available bonds and guarantees and, where appropriate, following the prescribed procedures (e.g. informing bondsmen).
7   Contacting the receivers and eliciting their views with regard to completion of the project.
8   Keeping a record of all time spent and costs incurred in dealing with and advising on the insolvency. It is normal for additional fees to be chargeable in this respect.

Completing the Works

There may be a number of options open to an Employer with regard to completing a project following the insolvency of a contractor. Surveyors should advise their clients of the alternative strategies available, together with the likely implications of each of these strategies, thus enabling their clients to make informed decisions. The alternatives available for completion may include:

Continuing with the original Contractor

This would also include reinstatement of the contractor's employment where determination had occurred. This is usually only practicable where a project is near completion and the contractor is able to continue.

Assignment and novation of the contract

If the receivers are able to sell the insolvent contracting organization, then assignment or novation of the contract with the purchasing contractor may be an attractive route for completing a project. *Assignment* is the legal term used to describe the transfer to a third party of a contractual right or obligation. The legal position is complex, and provision for assignment under JCT contracts is confusing. Novation is often preferred by both receivers and Employers. *Novation* involves forming a new agreement with a new contractor for completing the work.

Appointing a new contractor to complete the project

In some situations it may be prudent to appoint a new contractor. This choice may depend upon the stage of progress on a particular project. If a contract has just commenced prior to insolvency, it may be possible to approach the second tenderer. On the other hand, if a project is near completion, and in the absence of an indication from the receivers that assignment or novation may be possible, appointing a new contractor may be the only course of action that will ensure the completion of the project.

When faced with making, or advising on, such a decision, it is important to:

(a) consider all the factors that affect the decision and select the strategy that offers most benefits;
(b) act quickly to ensure that losses are minimized; and
(c) keep all options open until a formal agreement for completion has been made.

The basis of continuation contracts

Where a new contractor (or a series of contractors) is to be appointed to complete a project, it will be necessary to consider the basis of the contract and the documentation to be used. Generally, the options available will be the same as those for all projects, although the decision will be affected by:

(i)   the time available to prepare documentation and agree terms;
(ii)   the scope and amount of work involved;
(iii) the progress made by the original contractor;
(iv)   the documentation and contractual basis of the original contract; and
(v)   the need to obtain competitive tenders.

Where a project commenced shortly before contractor insolvency, it may be possible to use the original tender and contract documentation with an addendum to take account of the work completed. In this situation, it may also be possible to negotiate a contract with one of the original tenderers for the project, preferably on the basis of their original tender. It may be appropriate in other situations to let the work on the basis of a prime cost contract, particularly where the scope of work is small and time is critical. If the only work required is to make good defects, the appointment of jobbing contractors as and when required may prove effective. It is not necessary to let completion contracts on the same basis as the original contract. The specific circumstances pertinent at the time the completion contract is being considered should indicate the most appropriate contractual route.

Nominated sub-contractors

The effect of contractor insolvency

If the employment of the contractor is determined, then the employment of nominated sub-contractors is automatically determined. For many projects, the work of nominated sub-contractors is fundamental to the completion of the project. Where a nominated sub-contractor has not commenced work, the situation is usually straightforward in terms of the sub-contractor's involvement in the completion contract. However, where a nominated sub-contractor has commenced work, there may be problems. If the work, or the status of the work, of a sub-contractor is such that it would be most appropriate (for the completion of the project) for the sub-contractor to be involved in the continuation contract, then this involvement should be explored. There may, however, be some problems in practice. The sub-contractor may be owed money in respect of work undertaken and may, in the absence of payment, not be prepared to continue. Under the JCT 80 Conditions, the Employer is empowered, upon certain events, to make direct payments to nominated sub-contractors. This right under JCT 80 ceases, however, in the event of a contractor's liquidation (although the cessation of this right is questionable – refer to Stewart and Billingham[3]).

Following a contractor's insolvency, it would be prudent for the design team to contact all nominated sub-contractors in order to establish the status of their sub-contract works and to explore the possibility, and the issues in respect of, their involvement in a completion contract. A decision will have to be made (assuming a sub-contractor is prepared to continue) whether to use the same sub-contractor or to appoint a new one. If using the same sub-contractor, it may prove impossible to agree a price at the same level as the original sub-contract. However, there may still be more benefits, both of time and cost, in using the same sub-contractor at higher rates than in appointing a new sub-contractor.

Insolvency of a nominated sub-contractor

In the event of the insolvency of a nominated sub-contractor, and the subsequent determination of his employment, the architect (under JCT 80 Conditions) is required to re-nominate within a reasonable time. This further nomination should be in respect of completing the sub-contract Works, including making good any defective work (see pp. 102–3). The money payable to the new sub-contractor must be included in interim certificates and added to the cost of the Works.

If either an administrator or an administrative receiver is appointed, or a voluntary arrangement is made by the sub-contractor, a contractor has the right, subject to the written consent of the architect, to determine the employment of the sub-contractor. In such a situation where a sub-contractor or receiver is prepared to continue the sub-contract work then, providing the contractor and the architect are reasonably satisfied, this consent may be withheld.

It may be permissible, under clause 25.4.5, for the architect to grant an extension of time where re-nomination takes a considerable time. However, there is no express provision in clause 26 for the payment of loss and expense incurred by the contractor in respect of re-nomination.

The insolvency of a nominated sub-contractor will invariably affect the success of a project and will often give rise to disputes concerning extra time and extra cost. It is, therefore, in the interests of all persons involved in a project to ensure that the process of selection of nominated sub-contractors is as effective as possible in avoiding the likelihood of these problems arising.

References

1. NEWMAN, P., *Insolvency Explained* (London: RIBA Publications Ltd, 1992).
2. Chartered Institute of Building, *Code of Estimating Practice, 5th edn*, 1983.
3. STEWART, A. and BILLINGHAM, E., Hands off! Receiver, *Building*, 13 August 1993, pp. 16–17.

Bibliography

1. NEWMAN, P., *Insolvency Explained* (London: RIBA Publications Ltd, 1992).
2. STEWART, A. and BILLINGHAM, E., Hands off! Receiver, *Building*, 13 August 1993, pp. 16–17.
3. PHIPPS, M., Plugging those insolvency gaps, *Building*, 9 October 1992; pp. 30–31.

21

Dispute resolution

Introduction

Disputes are a common occurrence within the construction industry. One has only to examine reported legal cases to get a feel for the number of disputes. Of course, this is only the tip of the iceberg. Many disputes are resolved before they reach this stage, either on 'the courtroom steps' or through arbitration or by some other method of resolution.

Relationships within construction, between clients, contractors, sub-contractors and suppliers, are often adversarial. The risks associated with construction projects can be high, the process is complex and obligations are often onerous. The way demand is put on the industry through competitive tendering procedures can often increase adversity. The need for contractors to 'win' contracts can often mean that a project gets off to a bad start, with the contractor battling to ensure cost recovery.

Disputes are damaging both to the industry and to the clients of the industry. The costs associated with resolving disputes are often high, and the time involved can be very long. The overall impact of disputes on individual projects can jeopardize the objectives of all involved in them. A preventative approach would obviously be the best situation, that is, removing the causes of disputes altogether. However, such an environment could only be created by radically changing the processes, attitudes and structures that lead to dispute. It is unlikely in the short term (if at all) that such changes will be made. There will, therefore, continue to be a need for disputes to be resolved.

This chapter examines the various methods and approaches to resolving disputes in connection with construction projects. It covers litigation, arbitration, alternative dispute resolution and adjudication.

Litigation

Traditionally, disputes that could not be resolved directly by the parties involved in the dispute were either dealt with by the courts or referred to arbitration. *Litigation* involves one party issuing a writ against another. The parties are required to prepare a case in support of their argument, and this is heard in court. The mechanics of litigation are governed by statute. Decisions are made on the basis of the evidence presented, together with the application of relevant legislation and precedent. The process is often long and expensive, and can be damaging to business relationships. The role of the surveyor in litigation will normally be as a witness of fact or as an expert witness. Expert witnesses may be called by a party to a dispute to give expert evidence of opinion to support an argument.

Litigation is often an ineffective way of deciding disputes in the context of business relationships. Court proceedings are more concerned with deciding a winner and a loser to the dispute rather than establishing a compromise solution which reflects the business needs of all concerned. Litigation is adversarial and will tend to increase the conflict between parties to a dispute. Such an adversarial approach can damage a business relationship; in the context of construction, this can cause untold damage to the successful completion of an ongoing project.

Arbitration

Arbitration is the process whereby the parties refer any disputes arising in the course of a contract to an agreed third party to be resolved. Where reference to arbitration is in writing, the procedure is governed by the various Arbitration Acts. The Acts require arbitration to be carried out in a judicial manner, and at the conclusion of a hearing the Arbitrator must provide reasons for any award made. The decision of the Arbitrator in relation to findings of fact is conclusive; there is no route for appeal. However, appeals can be made to the courts on points of law, but only with the approval of the court hearing the appeal. The award of an Arbitrator is legally binding.

The advantages of using arbitration as opposed to litigation for resolving disputes include:

1 *Cost* – arbitration is generally much cheaper. However, the costs of arbitration are significant, and it is widely felt that the cost benefit has been eroded over time.
2 *Speed* – arbitration is generally much quicker.
3 *Flexibility* – the parties have more control over the timing and the format of the proceedings. Mutually agreeable arrangements can be made to cause minimum disturbance to normal business.
4 *Technical expertise of the arbitrator* – the Arbitrator will be selected by the parties on the basis of technical expertise and suitability to resolve the dispute in question. Decisions should, therefore, tend to be more in keeping with actual practice in the area under dispute. In litigation, a judge lacking detailed technical knowledge may be appointed.
5 *Privacy* – arbitration hearings are normally held behind closed doors. Sensitive business information is therefore kept private, and the harm that can result from damaging publicity is avoided.

Arbitration is the most commonly used formal method of resolving construction disputes out of court. Once the parties to a contract have agreed to settle a dispute by reference to arbitration, this is the method that should be used. Agreement to arbitration can be made at the time of dispute or by an agreement in the contract. In most situations the courts will refuse to hear a case where referral to arbitration has been agreed by the parties involved.

In their unamended state, the JCT Forms of Contract prescribe the use of arbitration. Article 5 of JCT 80 is the agreement requiring disputes between the Employer and the contractor to be referred to arbitration, and this should be effected in accordance with clause 41.

Clause 41.1 requires one of the parties to a dispute to give written notice to the other party to the effect that a dispute or difference will be referred to arbitration. The Arbitrator is to be a person agreed by the parties; if the parties cannot agree within 14 days of the written notice, then a person is to be appointed by the person named in the Appendix to the contract. The notice should identify:

(i)   the agreement to which the notice relates;
(ii)   the matters in dispute and to be referred to arbitration; and
(iii)   that appointment of an agreed Arbitrator is required, but in the absence of agreement the Appointor identified in the contract will be requested to appoint an Arbitrator.

Clause 41.2 provides that where a dispute involving nominated sub-contractors or nominated suppliers has been referred to arbitration, any related dispute between the contractor and the Employer should be referred to the same Arbitrator (joinder provisions). This requirement does not apply where either the Employer or the contractor considers that the appointed Arbitrator is inappropriately qualified in respect of the main contract dispute, or the Appendix has been constructed to exclude the operation of clauses 41.2.1 and 41.2.2.

Clause 41.3 precludes reference to arbitration prior to Practical Completion (or termination of the contractor's employment or abandonment of the Works) unless the parties agree. There are, however, a number of matters that can be referred to arbitration at any time; these are listed in clause 41.3.

Clause 41.4 gives wide powers to an Arbitrator to determine all matters in dispute. These powers include reviewing and revising certificates, opinions, decisions, requirements or notices. Any award of an Arbitrator is final and binding. However, appeals and applications can be made to the High Court in connection with questions of law. The provisions of the Arbitration Acts apply to arbitrations arising in connection with JCT contracts. Clause 41.9 prescribes that arbitration must be conducted in accordance with the JCT Arbitration Rules.

The JCT Arbitration Rules

The JCT Arbitration Rules are intended for use with JCT Standard Forms of Building Contract. The Rules describe the procedure and programme to be adopted in the event of arbitration. The party requiring a dispute to be referred to arbitration is termed the *Claimant*, the other party is the *Respondent*. There are three alternative procedures for the conduct of the arbitration; the parties elect which procedure to adopt. If the parties cannot agree on the procedure to adopt, then the Arbitrator has a duty to decide. The alternative procedures are:

1 Rule 5 – Procedure without hearing

From the time Rule 5 becomes applicable, the Claimant has 14 days in which to serve a statement of the case. Following submission of the Claimant's statement, the Respondent has 14 days in which to submit a statement of defence, together with any counterclaim. The Claimant has 14 days in which to serve a statement in reply to this defence, together with a defence to any counterclaims where applicable. If the Claimant serves a defence to a counterclaim, the Respondent has 14 days in which to serve a statement in reply to this defence. Any appropriate documents necessary to support statements should be identified by the Claimant and the Respondent in their statements; any documents fundamental to the claim or defence should be included with the statements. If, at any time in this programme, a party does not serve a statement within the required timescale the Arbitrator notifies the parties that he intends to proceed unless the relevant statement is received within 7 days of the notice. The Arbitrator should publish his award within 28 days. If the Claimant fails to serve a statement of the case, the Arbitrator's award will dismiss the claim and require the Claimant to pay the Arbitrator's fees and expenses and costs incurred by the Respondent. In some circumstances, the Arbitrator has the power to extend the timescales prescribed under Rule 5.

Rule 5 is suitable in situations where the Arbitrator is able to decide on the basis of written statements and there is no dispute over the facts.

2 Rule 6 – Full procedure with hearing

The procedure under Rule 6 is similar to that under Rule 5 concerning pleadings and discovery with the exception of time limits. The Claimant has 28 days in which to serve a statement of the case, after which the Respondent has 28 days in which to serve a defence and counterclaim. The Claimant subsequently has 14 days in which to serve a reply to the defence and 28 days in which to serve a defence to any counterclaims. If the Claimant serves a defence to a counterclaim, the Respondent has 14 days in which to serve a reply to the defence. Following this, the Arbitrator will consult with the parties and notify them of the time and place of the hearing. The Arbitrator should publish an award within 28 days of this hearing.

Rule 6 is appropriate where there is a disagreement between the parties on the facts associated with the dispute and the Arbitrator would require oral evidence, together with cross-examination, to establish the facts.

3   Rule 7 – Short procedure with hearing

The procedure requires a hearing to take place within 21 days of Rule 7 becoming applicable. At least 7 days before this hearing, any necessary supporting documentation should be identified by the parties and, where the documentation is not in the possession of one of the parties, it should be served upon that party. The documents identified should be issued to the Arbitrator at least 7 days prior to the hearing. At the end of the hearing, the Arbitrator can make an award to be subsequently supported in writing or publish the award within 7 days of the hearing. Each party to the dispute is required to bear its own costs (except for special reasons at the discretion of the Arbitrator).

The Rule 7 procedure is appropriate where there is no dispute of facts that would require cross-examination and where the dispute arises during the course of a project and a quick, binding award is required to enable the project to proceed.

The Arbitrator is empowered to inspect work or materials as necessary. In any award, he decides the liability of the parties in respect of the payment of arbitration fees and expenses and on liability for paying the other party's costs (except generally under the Rule 7 procedure). In making an award, an Arbitrator is not required to give reasons for the award unless required in writing to do so by one of the parties. The Arbitrator is required to notify the parties when an award is ready for publication, together with the amount of fees and expenses payable. On payment of the Arbitrator's fees and expenses, the award is normally taken up by the successful party.

Alternative dispute resolution

Alternative dispute resolution is the term used to describe a number of formal, non-adversarial methods of resolving disputes without resorting to the courts or arbitration. The approach was developed in the United States and was introduced to the UK in the 1980s. The processes offer an inexpensive, expeditious and probably more palatable alternative to litigation and arbitration. As with arbitration and litigation, the processes involve a third party. However, unlike arbitration and litigation, the role of the third party is not to sit in judgement, but to act as a neutral facilitator to a negotiated settlement between the parties.

The advantages of alternative dispute resolution are often referred to as the four Cs. These are:

1 *Consensus* – this approach requires the agreement of all parties to a resolution of the dispute. The emphasis is on finding a business, rather than a legal or adversarial, solution to the dispute.
2 *Continuity of business relations* – the processes are concerned with resolving disputes within the context of, and without permanently damaging, ongoing business relations.
3 *Control* – resolution of the dispute remains in the control of the parties to it. The parties can concentrate on forging a settlement which focuses on commercial issues rather than the letter of the law and may thus be less damaging for all parties. Once a dispute is referred to the courts or to arbitration, the parties effectively lose control of the process.
4 *Confidentiality* – the proceedings are not published, and therefore the damage resulting from adverse publicity is avoided.

One of the requirements of alternative dispute resolution is that it must be non-binding. If the process is not working, recourse to litigation or arbitration, as appropriate, must be available. If agreement is not reached, the process will seldom have been a waste of time, effort and expense. It will probably have clarified or narrowed the scope of a dispute and will undoubtedly reduce some of the information gathering and synthesis requirements of arbitration or litigation. Where an agreement is reached using one of the alternative methods of dispute resolution, it should be formalized into a written agreement between the parties. Some of the methods of alternative dispute resolution are described below.

Mediation

Under *mediation*, the parties in dispute select an independent third party to assist them in reaching an acceptable settlement to their dispute. This mediator should be skilled in problem solving and preferably have expertise relative to the dispute in question. The role of the mediator is not to make a judgement of the dispute, but to facilitate a settlement between the parties. The normal process involves the mediator meeting with the parties to agree the format and programme. The sessions usually begin jointly with each of the parties presenting their case, informally, to the mediator. This is followed by private sessions, known as 'caucuses', between the mediator and each of the parties. The mediator's role will be to get each of the parties to focus on their main interests and to get them to move to a common position. The mediator will move between caucuses, often passing offers from one party to the other. As he or she will be in a position of knowledge, confidentiality and impartiality are essential. For the same reason, the mediator is in a position to facilitate, and coax the parties into, the best possible negotiated settlement.

Conciliation

Conciliation is a similar process to mediation. The distinctions are that a conciliator will actively participate in the discussions between the parties, offering views on the cases put forward. There are no private meetings between the conciliator and individual parties to the dispute. It is more informal than mediation, perhaps aimed at getting the parties to discuss differences. Should the parties fail to reach agreement, it is common for the conciliator to recommend how the dispute should be settled.

Mini-trial

A tribunal, usually comprising senior management from the various parties and chaired by an independent adviser, hears presentations made by all parties to the dispute, who may be represented by lawyers. Witnesses and experts may be called to give evidence. Following the presentations, the senior management enter into negotiations, with the objective of reaching an agreed settlement. The chair's role is to facilitate negotiations, adding suggestions and advice as appropriate to encourage agreement. Should settlement not be reached, the chair may, depending upon the agreed procedure, offer a non-binding opinion. The constitution of the tribunal is important. Members of the tribunal should be senior management who have not had direct involvement with the project in dispute and possess the relevant authority for negotiating a business settlement. The chair should have the necessary technical and legal experience particular to the dispute, together with the skills necessary to facilitate a negotiated settlement.

Adjudication

Adjudication is a process aimed at providing a quick, non-binding solution to a dispute. The process involves an agreed third party giving a decision on disputes relating to a contract as and when they are referred to that third party. The adjudicator, together with the procedures and the types of dispute which can be referred to adjudication, will all be agreed and stated in the conditions of contract. For adjudication to be effective, any decision of the adjudicator must be immediately implemented. Appeal should only be permitted after Practical Completion has been achieved.

The process of adjudication provides a speedy resolution to a dispute, thus allowing the parties to proceed with minimal delay and damage to relationships and the project. It may also reduce the number of disputes taken to arbitration and litigation, as it may give a good indication of what any decision is likely to be using these methods of dispute resolution and therefore dissuade the parties from taking further action.

Conclusion

The use of alternative methods of dispute resolution and adjudication is not widespread in the UK, although the advantages these methods offer could benefit all involved in construction when compared with arbitration and litigation. JCT 80 contains provisions for disputes to be resolved by reference to arbitration, there being no provision for other formal approaches to dispute resolution. The recommendations of the Latham Report[1] are that all standard forms of contract should include provisions for adjudication, and that 'there should be no restrictions on the issues capable of being referred to the adjudicator, conciliator or mediator, either in the main contract or sub-contract documentation'. The Engineering and Construction Contract provides for the use of adjudication for resolving disputes. The ICE Minor Works Form makes provision for the use of conciliation procedures.

References

1. LATHAM, Sir Michael, *Constructing the Team* (London: HMSO, 1994), Chapter 9.

Bibliography

1. MARSHALL, E. A., *Gill: The Law of Arbitration* (London: Sweet and Maxwell, 1990).
2 LATHAM, Sir Michael, *Constructing the Team* (London: HMSO, 1994), Chapter 9.
3 MCINTYRE, J., Disputes under review, *Chartered Quantity Surveyor*, November 1991, pp. 16–17.
4 SHIFFER, R. A. and SHAPIRO, E., Introducing the Middlemen, *Contract Journal*, 5 April 1990, pp. 18–19.
5 BINGHAM, A., No losers when commerce triumphs over litigation, *Building*, 14 September 1990, p. 48.
6 DIXON, G., Finding a real alternative, *Chartered Quantity Surveyor*, February 1991, p. 20.
7 ASHWORTH, A., *Contractual Procedures in the Construction Industry*, 2nd edn (Essex: Longman Scientific & Technical, 1991), Chapter 28, Arbitration.

22

Contract selection

Introduction

It is normal, but not essential, for standard forms of contract to be used as the basis for an agreement between a developer and a contractor. The complex nature of construction, together with the high levels of risk often involved, usually mean that the use of a standard form of contract is advisable, certainly for all but the smallest of projects. Standard forms of contract offer the advantages of being carefully drafted by experts, being tried and tested and therefore 'known quantities', and being recognized by both the industry and the courts. Although such forms may have their individual weaknesses, it is often safer to rely on the 'devil you know'!

Surveyors must have a working knowledge of the availability and suitability of the various standard forms of construction contracts. Familiarity with one standard form of contract should not be a reason for surveyors always recommending the use of that form for all the projects they are involved with. It is imperative, if clients' interests are to be safeguarded, to recommend the form of contract most suited to each set of project circumstances. In this respect, it is incorrect to suggest that one form of contract is always better than another – the various standard forms have been drafted for use in specific circumstances. Using an inappropriate standard form of contract for a project is dangerous. It will often mean that objectives, in terms of time, cost and quality, are not fully realized and that the likelihood of disputes will increase. It is therefore essential for surveyors to develop effective contract selection skills.

This chapter identifies and explains the commonly used standard forms of construction contract. It considers the suitability of these forms for some of the various development situations. In conclusion, it offers a strategy for effective contract selection.

Standard forms of construction contract

The nature of standard forms of construction contract

Standard forms of construction contract have been developed to provide formal, predetermined arrangements and mechanisms to cope with the situations that can arise during the course of a construction project. They define the obligations and liabilities of the parties to the contract. They also allocate risks to the parties, different standard forms of contract generally allocating risks differently between contractor and Employer. Business is often described as the process of taking risk in return for reward. Contractors will accordingly price the risks that a contract requires them to bear, and this price will be included within tenders. It is not therefore, as sometimes thought, necessarily advantageous to transfer as much risk as possible to the contractor. Decisions should be made, as part of the procurement route and contract selection process, about which party is most suited to manage each of the individual risks. The chosen contract strategy for any project should reflect these decisions. Murdoch and Hughes[1] note that 'any decision about laying off risks on to others must involve weighing up the frequency of occurrence against the level of premium being paid for the transfer'. The choice of procurement and contract strategy should be that offering most value to a client, bearing in mind their objectives.

The 'family hierarchy' of contracts

The various standard forms of construction contract can be considered in terms of a 'family hierarchy'. The body responsible for drafting individual standard forms can be considered as the first level in the hierarchy of standard forms of construction contracts. So, for example, the contract examined in detail in this book, JCT 80, has been written by the Joint Contracts Tribunal for the Standard Form of Building Contract. JCT 80 is one form of contract which sits within the family of contracts produced by the Joint Contracts Tribunal. This family can be considered as the subsequent level within the hierarchy of standard forms of contract. Moving down the hierarchy from main forms of contract lie the various sub-contract forms, supplements, warranties and other ancillary documents.

Contract selection is, therefore, concerned with, first, selecting the appropriate family of contracts and, secondly, identifying the suitable main form and ancillary documents from within that family. The available families of standard contracts include the following:

1 *The Joint Contracts Tribunal (JCT)* – JCT contracts are considered by many as the 'industry standard'. They are certainly among the most comprehensive, and they are the most widely used standard forms of contract. The JCT family of contracts covers most forms of procurement and building types, and there is an impressive collection of ancillary documentation published to support the main forms. The JCT comprises interest groups from all sectors of the industry; its objective is to determine the format, content and wording of standard forms of contract. As such, JCT contracts are considered by many to be fair in that they are not loaded in favour of either party to a contract. They are, however, sometimes described as compromise conditions which, in trying to satisfy the interests of all, are unnecessarily long and complex. The variety of forms available from the JCT will be considered in more detail later in this chapter.

2 *Association of Consultant Architects' Form of Building Agreement* – The ACA form of contract was first published in 1982 by the Association of Consultant Architects. The latest edition, published in 1984, is ACA2. The form originated in response to the Association's dissatisfaction with JCT contracts, in particular, with the allocation of risks and responsibilities. Unlike JCT contracts, this form of contract has not been created by negotiation. The form is flexible in use; by appropriate selection of alternative clauses it can be used for both traditional and design and build procurement methods. However, the form is not widely accepted, and its use has been minimal.

3   *GC/Works Contracts* – General Conditions of Government Contracts for Building and Civil Engineering Works (GC/Works/1) is the form of contract often used for central government works. It is not a negotiated form of contract. GC/Works/2 is available for small or simple contracts. The forms are published by HMSO.

4   *The British Property Federation System* – The BPF System was introduced in 1983. The system presented radical changes to the way buildings are procured. The system adopts an amended version of the ACA2 form of contract. The system has received very little use.

5   *ASI Contracts* – The ASI Building Contract is produced by the Architects' and Surveyors' Institute. It suffers from the usual concerns of non-negotiated forms of contract in that some of the conditions may be considered as unfair contract terms. A standard form of sub-contract is published for use with the main form. Chappell[2] notes about the ASI Building Contract that 'it is intended for use on large projects, but one must doubt the wisdom of doing so in view of the ambiguous wording and inadequacy of some of the provisions'. The ASI also publishes Small Works and Minor Works contracts.

6   *ICE Contracts* – The ICE Contract is published by the Institution of Civil Engineers and has been prepared by the Institution of Civil Engineers in conjunction with the Association of Consulting Engineers and the Federation of Civil Engineering Contractors. It is a negotiated form of contract recommended for use on major civil engineering contracts. It can be used for public or private contracts. The sixth edition, ICE6, was published in 1991 and is a re-measurement contract; contractors are paid at contract rates on the basis of work undertaken. There is a standard form of bond to accompany the contract.

7   *FIDIC Conditions of Contract* – The Conditions, produced by the International Federation of Consulting Engineers, are intended for use internationally. They are broadly based upon the ICE Conditions.

8   *The Engineering and Construction Contract* – The first edition of the contract (called the New Engineering Contract) was published by the Institution of Civil Engineers in March 1993. Publication followed a review of alternative contract strategies for civil engineering work which was aimed at identifying

good practice. Following publication of Sir Michael Latham's report[3], in which adoption of the form was recommended, amendments were made to the contract and a second edition published in November 1995 with the new title – The Engineering and Construction Contract (ECC).

The objectives in drafting the form were threefold. First, to create a contract that was flexible in use. The ECC is comprehensive; it is designed to cater for all types of project, forms of procurement, methods of tender, and for use in any country. Secondly, to offer a contract which is clear and simple in terms of language and structure. Thirdly, to provide a contract which acts as a stimulus for good management.

At the time of writing, use of the ECC has not been widespread; the contract has, however, received good reviews from many within the industry, especially concerning the way it challenges many of the contemporary problems prominent within the construction industry. The contract offers a challenge to the JCT family of contracts, and it will be interesting to monitor how it affects their use.

The JCT family of contracts

The Joint Contracts Tribunal produces a wide range of standard forms of contract for use in different situations. They are most suitable for building, as opposed to engineering, projects. Surveys by the RICS have shown that JCT contracts are by far the most widely used standard forms of building contract in the UK. The main forms, together with some of the associated ancillary documents, are considered below.

1 *JCT 80 Standard Form of Building Contract* – There are three basic variants of JCT 80: 'with quantities', 'without quantities' and 'with approximate quantities'. All variants require Employers to appoint designers to carry out the design function.

    The *'with quantities'* edition is intended for use where the work is designed prior to contract and a bill of quantities has been prepared setting out the quality and quantity of the works. Contract documents comprise the form of contract, contract drawings, and contract bills. This is a lump sum form of contract; there is an agreed contract sum to be paid to the contractor by the Employer. Contractor's risk is limited to price only – the Employer bears the risk of errors in the bill of quantities.

The *'with approximate quantities'* standard form of contract enables construction to commence prior to the design being completed. In practice, a significant proportion of the design decisions must be made prior to contract. Approximate bills of quantities set out the quality and approximate quantity of work. This is a re-measurement form of agreement – the actual work required is measured and priced in accordance with the mechanisms and prices included in the agreement. There is no contract sum. The contract value is ascertained after the contract has been agreed.

The *'without quantities'* standard form of contract is intended for use where the design is completed prior to contract but where there is no bill of quantities. The contract documents will normally include drawings, specification, and a schedule of rates. The costs to contractors of tendering for 'without quantities' contracts are higher than for contracts based upon bills of quantities because contractors have to analyse the design and ascertain the quantities required themselves. For this reason, it is unusual, except perhaps in the case of negotiated contracts, to use this form for high-value projects. Contractor's risk includes both price and quantity. This is a lump sum form of contract.

Each of the three variants is available in two formats: for private sector or local authority use. Although these contracts can be used on projects of any size, it is unusual, bearing in mind the alternative forms available from the JCT, to use these forms for projects with a value below £200,000. The contracts are comprehensive in providing for many eventualities, but suffer from being complex in wording and numbering. The contracts are regularly amended and updated by the JCT to take into account developments in the law; at the time of writing the JCT had published fifteen amendments to the 1980 contract.

The following supporting documentation is published for use with JCT 80:

(i)   *Sectional Completion Supplement* – for use where a contract is to be completed in phases. The supplement sets down the agreement for phased completion and hand-over and contains the necessary consequential amendments to the standard form. The supplement is available in two formats: one for use with the 'with quantities' and 'with approximate quantities' forms and the other for use with the 'without quantities' form.

(ii) *Contractor's Designed Portion Supplement* – for use where the contractor is required to undertake part of the design (note that this is not a design and build contract).

(iii) *Nominated Sub-contract documentation* – for use where sub-contractors are nominated in accordance with the procedures prescribed in clause 35 of JCT 80. The documentation is as follows:
   – *NSC/T* – for use for nominated sub-contract tendering;
   – *NSC/A* – the sub-contract agreement between a contractor and a nominated sub-contractor;
   – *NSC/N* – for use for a nomination instruction;
   – *NSC/W* – the warranty agreement between a nominated sub-contractor and an Employer;
   – *NSC/C* – the conditions of nominated sub-contract.

(iv) *Domestic Sub-contract documentation* – forms DOM/1, the Articles of Agreement, and DOM/1c, the Conditions of Sub-contract, are for use where the contractor appoints domestic (not nominated) sub-contractors.

(v) *Nominated Supplier tender and warranty forms* – forms TNS/1 and TNS/2 can be used for obtaining tenders from nominated suppliers.

(vi) *Fluctuation clauses* – separate conditions for use with private and local authority versions of the main form are available.

2 *JCT Agreement for Minor Building Works (MW 80)* – MW 80 is designed for use on projects which are short in duration, small and simple. The contract is short and easy to follow. The same form can be used for both private employers and local authorities. MW 80 is a lump sum form of contract; design should be completed by the Employer's representative (usually an architect or building surveyor) prior to a contract being agreed. There is no provision in the form for the use of bills of quantities; contract documents will usually include

a combination of drawings, specifications and schedules of works. Practice Note M2 suggests that the form is suitable for contracts with a value of up to £70,000 (at 1987 price levels); this is a guide only, and MW 80 has been used successfully on contracts with a value far greater than the recommended limit. The disadvantages of the form – or perhaps the reason why it is only intended for small, simple projects – lie in its lack of comprehensiveness. The provisions dealing with claims are limited, and there are no provisions for fluctuations, nominated sub-contractors or nominated suppliers.

3   *JCT Intermediate Form of Building Contract (IFC 84)* – IFC 84 was introduced by the JCT to fill the gap between the complex JCT 80 Contract and the simpler MW 80. The form is suitable for projects which are simple in content, require only basic trades and skills, where the services installations are not complex, there are no specialist installations, and the works have been designed by the Employer prior to tender. As a guide, it is recommended that the contract is used where the contract period does not exceed one year and the contract value is below £280,000 (at 1987 price levels). IFC 84 can be used for both private employers and local authorities. It is flexible in terms of the documentation that may be used in conjunction with it; together with drawings, the form permits the use of specifications, or schedules of work, or bills of quantities. There are no provisions for the nomination of sub-contractors, but the form permits sub-contractors to be named by the Employer. The following supporting documentation is produced for use with IFC 84:

(i)   *Named Sub-contract documentation* – NAM/T is the form of tender and agreement and NAM/SC are the conditions of sub-contract for use when there are named sub-contractors. ESA/1 is an agreement for use between the Employer and a sub-contractor in situations where the sub-contractor has a design responsibility.

(ii)   *Sectional completion supplement* – for use where a project is required to be completed in phases.

(iii)   *Fluctuations clauses.*

4   *JCT Standard Form of Building Contract with Contractor's Design (CD 81)* – CD 81 is intended to be used when the contractor is responsible for design and construction. It is

similar in content and complexity to JCT 80. The contract is flexible in that it caters for both private and local authority employers and that it permits design input by the Employer up to tender stage. There is no mention of architect or quantity surveyor in the contract; contract administration is performed by a duty holder referred to as the Employer's agent (in practice, the role of Employer's agent could be undertaken by a surveyor, an architect, an engineer, a project manager, or a combination of these disciplines).

In essence, the contract operates such that tenders are invited on the basis of satisfying a set of Employer's Requirements. The nature of the Employer's Requirements can, in practice, vary from brief performance requirements to detailed designs. Tenders are submitted in the form of Contractor's Proposals (the contractor's response to the Employer's Requirements). These subsequently form the basis for the contract.

The form received immediate, widespread use on its introduction, and its popularity has continued to grow. The form provides employers with the reassurance of single point responsibility for design and construction. It provides a low-risk arrangement for employers so long as they comply with the spirit of the form and keep post-contract variations to the Contractor's Proposals to a minimum. The reasons for popularity may also include effective marketing of the design and build process by design and build contractors; and dissatisfaction from employers with the unpredictability of performance using other procurement and contractual arrangements.

5 *JCT Management Contract (MC 87)* – Under MC 87 a Management Contractor is appointed to arrange and co-ordinate construction work. The Management Contractor does not carry out any of the work; this is undertaken by a series of Works Contractors. The Employer is required to arrange for the design to be carried out by a design team. The Management Contractor acts as an interface between the designers and the Works Contractors. He is paid on the basis of a fee for managing the Works, together with the actual cost (or prime cost) of the Works Contracts plus the cost of providing site facilities. The Employer's financial commitment will not be known before work commences, and probably not until a project is nearing or has reached completion – effective cost

management and forecasting is therefore fundamental in achieving an Employer's financial objectives. Employers bear much of the financial risk under this form of arrangement. The contract does, however, offer the opportunity of commencing work on site prior to the design being completed. This should mean an overall reduction of the development programme for a project when compared with traditional procurement approaches. Other advantages include:

(i)   the ability to let Works Contracts on the basis of competition, which should ensure that the overall project cost is representative of market prices;

(ii)   in a falling market, Works Contracts will be representative of the falling market (they are normally progressively let as the programme requires) and thus the Employer should receive any associated financial benefits;

(iii)   in a rising market, Works Contracts will be representative of the rising market, and perhaps the likelihood of contractor insolvency, and the detrimental effect of insolvency on the success of a project, will be reduced;

(iv)   contractor input can only add to the efficiency of designs;

(v)   ability to alter the project requirements during construction.

The benefits offered by management contracts are greatest where projects are high in value, long in duration, and complex. The JCT produces the following documentation for use in conjunction with MC 87:

−*Works Contract/1* − comprising: Section 1, Invitation to Tender; Section 2, Works Contractor's Tender; and Section 3, Articles of Agreement Attestation.

−*Works Contract/2* − the conditions of contract between the Management Contractor and a Works Contractor.

−*Works Contract/3* − an agreement between the Employer and a Works Contractor.

−*Phased completion supplement.*

6   *JCT Standard Form of Prime Cost Contract (PCC 92)* − PCC 92 was introduced to replace the Fixed Fee Form of Prime Cost Contract. It is broadly based upon the provisions of JCT 80. The contractor is paid the cost of carrying out the work plus a fee to cover profit and overheads. The Employer is responsible for

arranging the design. Although the design does not have to be completed prior to work starting on site, the design must be developed sufficiently to enable the contractor to ascertain the appropriate fee. The fee can be either fixed or paid on the basis of a percentage of the prime cost. Where the fee is fixed, the contract provides for adjustment of the fee if the difference between the actual prime cost and the estimated prime cost is greater than 10 per cent. The Employer bears the majority of financial risk under this form of contract. The total cost of a project will not be known until completion.

PCC 92 should be used in situations where it is not possible to define the extent of the works prior to tender, either due to lack of time or because the nature of the work is such that it cannot be defined (for example, some types of repair and refurbishment work).

7   *JCT Measured Term Contract (JCT 89)* – The measured term contract is intended to be used where an Employer requires maintenance or minor works to be undertaken on a regular basis. Contracts are usually agreed on the basis of a schedule of rates for carrying out certain types of work on a defined list of properties. The rates are used to value the work carried out over a defined period of time. Contracts can be let on a fixed or fluctuating price basis. The standard form contains a break provision for terminating the contract early.

An approach to effective contract selection

The governing principles

The requirements and objectives of the Employer will be the primary factors driving contract selection. For most projects, the selection of procurement route to meet the Employer's needs will limit the available choices of standard form to be considered. In most situations, procurement and contract selection will be considered simultaneously.

The decision about which form of contract to use for a project will be based upon the following factors:

1   *The nature of the client* – Answers to the following questions may be pertinent to contract selection:

(i)   Are clients private individuals or organizations or public bodies?

(ii) Are clients regular/frequent developers or is a one-off project envisaged?

(iii) How knowledgeable are clients about their needs?

(iv) How knowledgeable are clients about the processes of construction and development?

(v)   What past experience of construction have clients had, and how does their experience prejudice the use of the various alternative procurement and contract choices?

2   *The risk attitude of the client* – What risks are individual clients prepared to take in relation to the associated rewards available? This will depend to some extent on the nature of the client. Public clients may tend to be more averse to risk than private clients. The nature of development may also affect risk attitude – clients engaging in speculative commercial developments may be prepared to take greater risks with regard to the construction cost of a project in order to receive the benefits associated with achieving earlier completion.

3   *The procurement method adopted* – Some of the standard forms of construction contract are written for use in connection with specific procurement strategies while others (for example, ECC) are flexible. The source of design, whether generated by the Employer or contractor, will also affect the choice of contract.

4   *The client's priorities in terms of time, cost and quality* – The balance of requirements will affect the choice of contract.

➤   If clients want cost certainty prior to agreeing a contract, then a lump sum form of contract is appropriate. This approach requires Employers either to allow sufficient time for their consultants to complete the design prior to a contract being agreed or to adopt a design and build approach. The first option will generally increase the overall development time, whereas the design and build approach may mean a reduction in quality.

If time is of the essence, then either a management contract, a prime cost contract or an approximate quantities contract may be appropriate, depending on the other decision factors. None of these forms of contract offer cost certainty.

Where quality is the main priority, the use of a design and build contract will probably be inappropriate.

5 *The size of the project* – The various types of JCT contract suitable for different sizes of project have been discussed above.

6 *The type of documentation being used* – Although contract selection will normally drive the requirement for contract documentation.

7 *The type of project* – Answers to the following questions will affect the decision on form of contract:
  (i)   is the project engineering or building based?
  (ii)   how complex is the project?
  (iii) does the project involve specialist installations or construction techniques?

In advising on contract strategy, it is important for surveyors to remain impartial. In many situations, the benefits – for example, higher fees – may be greater under one contract strategy than another. This must not influence the advice given. The recommended contract should be that which offers the greatest value to a client. Conversely, it may be necessary in some situations to advise clients that, although they have been offered a service by, for example, a design and build contractor that purports to offer the best solution to their needs, this may not be the case. Of course, it is in the design and build contractor's interest to sell his service.

It is often the case that a bad experience in using a particular form of contract can prejudice both a client and a consultant against using or recommending that form again. Problems and difficulties can occur under any form of contract, and it is probable that the problem is not down to the contract but is caused by some other factor. Bad experience should not necessarily preclude the use of a standard form of contract on future projects.

A strategy for selection

The wide range of available standard contract options makes the use of decision-making techniques advisable to ensure that rational selections are made. The experienced practitioner will often develop an intuitive skill in selecting an appropriate form of contract. However, for the inexperienced, a more formal approach to selection is often helpful. Indeed, where a client requires a detailed account of the reasons behind a recommended contract strategy, it is sometimes helpful to adopt a formal approach to explaining a particular rationale.

The criteria for decision-making will generally flow from the requirements of the Employer, and to ensure effective decision-making it is essential to define these criteria. There will be a difference between the wants and needs of an Employer. While it is not unreasonable for an Employer to want the best of all worlds in terms of low cost, low risk, short programme, high quality, cost certainty, etc., it is not usually possible for procurement and contract strategy to provide this. It is therefore necessary to identify Employer needs. This can be done by prioritizing those wants which are essential to the Employer. Having established this hierarchy of needs, it should be possible, in conjunction with the other decision-making criteria, such as the size and complexity of the project, to identify the contract strategy that is most suitable for each given situation. An approach similar to that described in Chapters 2 and 3 in connection with procurement is appropriate.

The selection of contract form will normally involve identifying the most appropriate family of contracts, selecting the individual form of contract to be used, and identifying the appropriate support documentation required. This chapter has briefly described the nature of the different families of contracts together with the main forms within the JCT family of contracts. This should help in directing the surveyor to an appropriate form for a given situation. However, it will be essential for the surveyor to analyse in detail the provisions within that form prior to making a recommendation. The aspects of the form which do not match the requirements of the project should be identified. If these are minor points, then the form of contract may still be suitable, perhaps with some minor amendment or addition (but note that any changes to standard forms of contract should be made by legal experts and preferably be limited to minor changes – see comment on p. 2). If the issues are major, an alternative contract strategy should be considered. Despite the wide choice of standard contracts available, it is unlikely that any contract will exactly match all the requirements of a project; therefore, sensible compromise will often be necessary.

In giving advice on contract selection, surveyors should clarify the reasons for their recommendations, together with the conditions that must be observed in consequence of the selection. Too many projects suffer as a result of the requirements pertaining to the use of a particular contract not being met. For example, if a project is being undertaken on the basis of a JCT 80 'without quantities' contract, the contract envisages that the design should be practically complete prior to the contract being agreed. However, in practice it is not uncommon to witness a lump sum contract being awarded when much of the design is incomplete. This can create risk for Employers, and is unnecessary in view of the numerous alternative methods of procurement and forms of contract which are designed to be used in situations where the work is to start before the design is complete.

References

1. MURDOCH, J. and HUGHES, W., *Construction Contracts Law and Management* (London: E & FN Spon, 1992), p. 15.
2. CHAPPELL, D., *Which Form of Building Contract?* (London: Architecture Design and Technology Press, 1991), p. 62.
3. LATHAM, Sir Michael, *Constructing the Team* (London: HMSO, 1994).

Bibliography

1. CHAPPELL, D., *Which Form of Building Contract?* (London: Architecture Design and Technology Press, 1991).
2. JCT, *Practice Note 20: Deciding on the appropriate form of JCT Main Contract* (London: RIBA Publications Ltd, 1993).
3. JCT, *Practice Note 7: Standard Form of Building Contract for use with bills of approximate quantities* (London: RIBA Publications Ltd, 1987).
4. ASHWORTH, A., *Contractual Procedures in the Construction Industry,* 2nd edn (Essex: Longman Scientific and Technical, 1991).
5. MURDOCH, J. and HUGHES, W., *Construction Contracts Law and Management* (London: E & FN Spon, 1992).
6. PIKE, A., *Practical Building Forms and Agreements* (London: E & FN Spon, 1993).

Appendix A

Particulars of contract referred to in the examples given in the text

Location of site: 28–34, Thames Street, Skinton

Description of works: Five lock-up shops at ground level
                                     with ten flats over and basement car
                                     park and stores

Contract sum: £495 000

Employer: Skinton Development Co., High Path,
                                     Skinton

Main contractor: Beecon Ltd., River Road, Skinton

Architect: Draw & Partners, 25 Bridge Street,
                                     Skinton

Quantity surveyor: Kewess & Partners, 52 High Street,
                                     Urbiston

Basis of contract: JCT Standard Form of Contract,
                                     Private with Quantities, 1980 Edition
                                     with Amendments

Clauses from the Conditions
of Contract not applicable: 22B, 22C

Entries on the Appendix to
the Conditions of Contract: *See* pp. 347–9

Commencing date: 21 March 1994

Completion date: 20 November 1995

Contract Period: 20 months

Abstract of information from Contract Bills

1. *Amounts inserted in Bill No. 1 – Preliminaries*

	£
Management and staff	20 000
Site accommodation	1 080
Electric lighting and power	180
Water	1 377
Telephones	450
Safety, health and welfare	3 600
Cleaning	540
Drying out	450
Security	1 350
Small plant and tools	3 050
Setting out	540
Temporary roads	1 350
Scaffolding	1 450
Fencing and hoardings	1 620
Total	35 660

2. *P.c. sums and profit and attendance amounts*

	£	£
Nominated sub-contractors:		
Piling	11 700	
Profit 5%	585	
Special attendance	450	
Asphalt tanking and roofing	16 200	
Profit 5%	810	
Special attendance	180	
Electrical installation	18 000	
Profit 5%	900	
Special attendance	450	
Lifts installation	34 200	
Profit 5%	1 710	
Special attendance	720	
Carried forward		85 905

	£	£
brought forward		85 905
Nominated suppliers:		
Kitchen fitments	3 600	
Profit 5%	180	
Ironmongery	2 700	
Profit 5%	135	
Sanitary appliances	4 320	
Profit 5 %	216	
		11 151

3. *Provisional sums*

	£	£
Statutory authorities:		
Water main connection	360	
Electricity mains connection	360	
Gas main connection	540	
Sewer connection	900	
		2 160
Dayworks:		
Labour	1 500	
Addition to net cost of labour 50%	750	
Materials	750	
Addition to net cost of materials 20%	150	
Plant	750	
Addition to net cost of plant 20%	150	
		4 050
Employer's telephone call charges	750	
Security installation	2 000	
External light fittings	1 558	
Landscaping	2 150	
Contingencies	8 000	
		14 458
Total		117 724

4. *Summary of Bill No. 3*

BILL NO 3 – SHOPS AND FLATS

SUMMARY

PAGE		£	p
26	GROUNDWORKS	31 256	57
33	IN SITU CONCRETE	54 488	22
44	MASONRY	45 417	33
69	STRUCTURAL/CARCASSING METAL/TIMBER	20 342	64
79	CLADDING/COVERING	9 245	12
92	WATERPROOFING	18 587	66
104	WINDOWS/DOORS/STAIRS	36 216	50
109	SURFACE FINISHES	15 100	50
112	FURNITURE/EQUIPMENT	3 819	96
116	BUILDING FABRIC SUNDRIES	9 256	15
124	DISPOSAL SYSTEMS	8 326	16
129	PIPED SUPPLY SYSTEMS	19 149	60
136	ELECTRICAL SERVICES	9 389	72
	TOTAL CARRIED TO GENERAL SUMMARY	£ 280 596	13

137

5. *General Summary*

GENERAL SUMMARY

PAGE		£	p
15	BILL NO 1		
	PRELIMINARIES/GENERAL CONDITIONS	35 660	00
19	BILL NO 2		
	DEMOLITION	5 531	00
137	BILL NO 3		
	SHOPS AND FLATS	280 596	13
143	BILL NO 4		
	EXTERNAL WORKS	40 556	51
147	BILL NO 5		
	PRIME COST AND PROVISIONAL SUMS	117 724	00
		480 067	64

Add for:

	£	p
Insurance against injury to persons and property as item 5C	3 632	00
All risks insurance as item 5C	811	36
Water for the Works as item 13A	1 377	00
	485 888	00
Add to adjust for errors	9 112	00
TOTAL CARRIED TO FORM OF TENDER	495 000	00

148

6. *List of basic prices of materials*

Basic prices of materials

The contractor is required to enter below the materials and prices to which he wishes clauses 39.3 and 39.4 of the Conditions of the Contract to apply. Only materials which are to be incorporated in the Works are to be included in the list and the prices are to be those that have been used in the pricing of the bills of quantities. The contractor shall furnish to the architect, upon request, all suppliers' quotations and invoices in support of the prices on the list and of any claim based thereon.

Claims will be accepted only where they relate to fluctuations in the market price of any material included in the list. Claims will not be accepted which are based upon variations in price which are due to purchasing in small quantities or to materials being of a different quality from that specified in the bills of quantities.

Materials	Unit	Prices
Ordinary Portland cement	tonne	£ 77.00
Fletton bricks	1 000	£100.00
Sawn softwood – 25 × 50 mm	100 m	£ 32.00
38 × 100 mm	100 m	£ 75.00
50 × 150 mm	100 m	£123.00

Appendix B

Articles of Agreement

made the ___*23*___ day of ___*February*___ 19 *94*

BETWEEN ___*A. Brown*___

of (or whose registered office is situated at) ___*Skinton Development*___

Company, High Path, Skinton, Middlesex

(hereinafter called 'the Employer') of the one part

AND ___*J. E. Green*___

of (or whose registered office is situated at) ___*Beecon Ltd,*___

River Road, Skinton, Middlesex

(hereinafter called 'the Contractor') (a) of the other part.

Footnote

(a) Where the Contractor is not a limited liability
company incorporated under the Companies Acts, see
Footnote (v) to clause 35·13·5·3·4.

Whereas

Recitals

First the Employer is desirous of [b] *erecting a three-storey*

block of shops and flats

at *28 - 34, Thames Street,*

Skinton, Middlesex,

and has caused Drawings and Bills of Quantities showing and describing the work to be done to be prepared by or under the direction of

1. Draw R I B A

Second the Contractor has supplied the Employer with a fully priced copy of the said Bills of Quantities (which copy is hereinafter referred to as 'the Contract Bills');

Third the said Drawings numbered *462/1 to 462/23*

(hereinafter referred to as 'the Contract Drawings') and the Contract Bills have been signed by or on behalf of the parties hereto;

Fourth the status of the Employer, for the purposes of the statutory tax deduction scheme under the Finance (No. 2) Act, 1975, as at the Base Date is stated in the Appendix;

[b·1] Fifth the extent of the application of the Construction (Design and Management) Regulations 1994 (the 'CDM Regulations') to the work referred to in the First recital is stated in the Appendix;

Footnotes [b] State nature of intended works.

[b·1] See the notes on the Fifth recital in Practice Note 27 'The application of the Construction (Design and Management) Regulations 1994 to Contracts on JCT Standard Forms of Contract' for the statutory obligations which must have been fulfilled before the Contractor can begin carrying out the Works.

6

P With 11/94

Now it is hereby agreed as follows

Contractor's obligations

Article 1

For the consideration hereinafter mentioned the Contractor will upon and subject to the Contract Documents carry out and complete the Works shown upon, described by or referred to in those Documents.

Contract Sum

Article 2

The Employer will pay to the Contractor the sum of ___ *four hundred*

and ninety-five thousand pounds

(£ *495,000* . *00*)

(hereinafter referred to as 'the Contract Sum') or such other sum as shall become payable hereunder at the times and in the manner specified in the Conditions.

Architect

Article 3

The term 'the Architect' in the Conditions shall mean the said

1. Draw R /BA

of ___ *25 Ridge Street, Skinton, Middlesex,*

or, in the event of his death or ceasing to be the Architect for the purpose of this Contract, such other person as the Employer shall nominate within a reasonable time but in any case no later than 21 days after such death or cessation for that purpose, not being a person to whom the Contractor no later than 7 days after such nomination shall object for reasons considered to be sufficient by an Arbitrator appointed in accordance with article 5. Provided always that no person subsequently appointed to be the Architect under this Contract shall be entitled to disregard or overrule any certificate or opinion or decision or approval or instruction given or expressed by the Architect for the time being.

Footnotes   [c] [d] [e] [f] Not used.

**Quantity
Surveyor**

Article 4
The term 'the Quantity Surveyor' in the Conditions shall mean

<u>C Kewess, FRICS</u>

of <u>52 High Street, Urkiston, Surrey</u>

or, in the event of his death or ceasing to be the Quantity Surveyor for the purpose of this Contract, such other person as the Employer shall nominate within a reasonable time but in any case no later than 21 days after such death or cessation for that purpose, not being a person to whom the Contractor no later than 7 days after such nomination shall object for reasons considered to be sufficient by an Arbitrator appointed in accordance with article 5.

**Settlement of
disputes –
Arbitration**

Article 5
If any dispute or difference as to the construction of this Contract or any matter or thing of whatsoever nature arising thereunder or in connection therewith shall arise between the Employer or the Architect on his behalf and the Contractor either during the progress or after the completion or abandonment of the Works or after the determination of the employment of the Contractor, except under clause 31 *(statutory tax deduction scheme)* to the extent provided in clause 31·9 or under clause 3 of the VAT Agreement, it shall be and is hereby referred to arbitration in accordance with clause 41.

**Planning
Supervisor [g·1]**

Article 6·1
The term 'the Planning Supervisor' in the Conditions shall mean the Architect

or

of

or in the event of the death of the Planning Supervisor or his ceasing to be the Planning Supervisor such other person as the Employer shall appoint as the Planning Supervisor pursuant to regulation 6(5) of the CDM Regulations.

**Principal
Contractor [g·1]**

Article 6·2
The term 'the Principal Contractor' in the Conditions shall mean the Contractor, or, in the event of his ceasing to be the Principal Contractor, such other contractor as the Employer shall appoint as the Principal Contractor pursuant to regulation 6(5) of the CDM Regulations.

Footnote

[g] Not used.

[g·1] Delete Articles 6·1 and 6·2 when only regulations 7 and 13 of the CDM Regulations apply; see Appendix under the reference to the Fifth recital.

P With 11/94

Notes

[A1] For Agreement executed under hand and NOT as a deed

[A1] **AS WITNESS THE HANDS OF THE PARTIES HERETO**

[A1] Signed by or on behalf of the Employer _____ *A. Brown*

in the presence of:

L. Simpson

[A1] Signed by or on behalf of the Contractor _____ *Jt. Green*

in the presence of:

L. Simpson

[A2] For Agreement executed as a deed under the law of England and Wales by a company or other body corporate: insert the name of the party mentioned and identified on page 1 and then use *either* [A3] and [A4] *or* [A5]. If the party is an *individual* see note [A6].

[A3] For use if the party is using its common seal, which should be affixed under the party's name.

[A4] For use of the party's officers authorised to affix its common seal.

[A2] **EXECUTED AS A DEED BY THE EMPLOYER**
hereinbefore mentioned namely _____

[A3]   by affixing hereto its common seal

[A4]   in the presence of:

* OR —

[A5] For use if the party is a company registered under the Companies Acts which is not using a common seal: insert the names of the two officers by whom the company is acting *who MUST be either a director and the company secretary or two directors*, and insert their signatures with 'Director' or 'Secretary' as appropriate. *This method of execution is NOT valid for local authorities or certain other bodies incorporated by Act of Parliament or by charter if exempted under s.718(2) of the Companies Act 1985.*

[A5]   acting by a director and its secretary*/two directors* whose signatures are here subscribed:
namely _____

[Signature]_____ *DIRECTOR*

and _____

[Signature]_____ *SECRETARY*/DIRECTOR**

[A2] **AND AS A DEED BY THE CONTRACTOR**
hereinbefore mentioned namely _____

[A3]   by affixing hereto its common seal

[A4]   in the presence of:

[A6] If executed as a deed by an *individual:* insert the name at [A2], delete the words at [A3], substitute 'whose signature is here subscribed' and insert the individual's signature. The individual MUST sign in the presence of a witness who attests the signature. Insert at [A4] the signature and name of the witness. Sealing by an individual is not required.

Other attestation clauses are required under the law of Scotland.

* OR —

[A5]   acting by a director and its secretary*/two directors* whose signatures are here subscribed:
namely _____

[Signature]_____ *DIRECTOR*

and _____

[Signature]_____ *SECRETARY*/DIRECTOR**

Appendix

Clause etc.

Statutory tax deduction scheme – Finance (No. 2) Act 1975	Fourth recital and 31	Employer at Base Date *~~is a 'contractor'~~/ is not a 'contractor' for the purposes of the Act and the Regulations
Base Date	1·3	<u>2 February 1994</u>
Date for Completion	1·3	<u>20 November 1995</u>
VAT Agreement	15·2	Clause 1A of the VAT Agreement *~~applies~~/does not apply [k·1]
Defects Liability Period (if none other stated is 6 months from the day named in the Certificate of Practical Completion of the Works)	17·2	
Assignment by Employer of benefits after Practical Completion	19·1·2	Clause 19·1·2 *~~applies~~/does not apply
Insurance cover for any one occurrence or series of occurrences arising out of one event	21·1·1	£ <u>1,500,000</u>
Insurance – liability of Employer	21·2·1	Insurance *~~may be required~~/ is not required Amount of indemnity for any one occurrence or series of occurrences arising out of one event £ _____ [y·1]
Insurance of the Works – alternative clauses	22·1	*Clause 22A/~~Clause 22B/ Clause 22C~~ applies (See Footnote [m] to clause 22)
Percentage to cover professional fees	*22A 22B·1 22C·2	<u>Ten per cent</u>
Annual renewal date of insurance as supplied by Contractor	22A·3·1	<u>15 January</u>

Footnotes

* Delete as applicable.

[k·1] Clause 1A can only apply where the Contractor is satisfied at the date the Contract is entered into that his output tax on all supplies to the Employer under the Contract will be at either a positive or a zero rate of tax.

On and from 1 April 1989 the supply in respect of a building designed for a 'relevant charitable purpose' (as defined in the legislation which gives statutory effect to the VAT changes operative from 1 April 1989) is only zero rated if the person to whom the supply is made has given to the Contractor a certificate in statutory form: see the VAT leaflet 708 revised 1989. Where a contract supply is zero rated by certificate only the person holding the certificate (usually the Contractor) may zero rate his supply.

This footnote repeats footnote [k·1] for clause 15·2.

[y·1] If the indemnity is to be for an aggregate amount and not for any one occurrence or series of occurrences the entry should make this clear.

Clause etc.

Insurance for Employer's loss of liquidated damages – clause 25·4·3	22D	Insurance ~~*may be required/~~ is not required
	22D·2	Period of time
		_____
Date of Possession	23·1·1	*21 March 1994*
Deferment of the Date of Possession	23·1·2 25·4·13 26·1	Clause 23·1·2 *applies/does not apply
		Period of deferment if it is to be less than 6 weeks is
		_____
Liquidated and ascertained damages	24·2	at the rate of
		£ *1,000* per *week*
Period of suspension (If none stated is 1 month)	28·2·2	_____
Period of delay (if none stated is, in respect of clauses 28A·1·1·1 to 28A·1·1·3, 3 months, and, in respect of clauses 28A·1·1·4 to 28A·1·1·6, 1 month)	28A·1·1·1 to 28A·1·1·3 28A·1·1·4 to 28A·1·1·6	_____ _____
Period of Interim Certificates (if none stated is 1 month)	30·1·3	_____
Retention Percentage (if less than 5 per cent) [aa]	30·4·1·1	*three per cent*
Work reserved for Nominated Sub-Contractors for which the Contractor desires to tender	35·2	_____
Fluctuations: (If alternative required is not shown clause 38 shall apply)	37	~~clause 38~~ [cc] clause 39 ~~clause 40~~
Percentage addition	38·7 or 39·8	*ten per cent*
Formula Rules	40·1·1·1	
	rule 3	Base Month
		_____ 19 _____
	rules 10 and 30 (i)	Part I/Part II [dd] of Section 2 of the Formula Rules is to apply

Footnotes

* Delete as applicable.

[z] Not used.

[aa] The percentage will be 5 per cent unless a lower rate is specified here.

[bb] Not used.

[cc] Delete alternatives not used.

[dd] Strike out according to which method of formula adjustment (Part I – Work Category Method or Part II – Work Group Method) has been stated in the documents issued to tenderers.

Clause etc.

Settlement of disputes – Arbitration – appointor (if no appointor is selected the appointor shall be the President or a Vice-President, Royal Institute of British Architects)	41·1	President or a Vice-President: *Royal Institute of British Architects *Royal Institution of Chartered Surveyors *Chartered Institute of Arbitrators
Settlement of disputes – Arbitration	41·2	Clauses 41·2·1 and 41·2·2 apply (See clause 41·2·3)
Performance Specified Work	42·1·1	Identify below or on a separate sheet each item of Performance Specified Work to be provided by the Contractor and insert the relevant reference in the Contract Bills [••]

Index to references to clauses in standard forms of contract

Subject Index